Sustainable Aviation

Thomas Walker · Angela Stefania Bergantino ·
Northrop Sprung-Much · Luisa Loiacono
Editors

Sustainable Aviation

Greening the Flight Path

palgrave
macmillan

Editors
Thomas Walker
Department of Finance
Concordia University
Montreal, QC, Canada

Northrop Sprung-Much
Concordia University
Montreal, QC, Canada

Angela Stefania Bergantino
Department of Economics,
Management and Business Law
University of Bari
Bari, Italy

Luisa Loiacono
University of Ferrara
Ferrara, Italy

ISBN 978-3-030-28660-6 ISBN 978-3-030-28661-3 (eBook)
https://doi.org/10.1007/978-3-030-28661-3

Cover credit: Tjasa Krievek

This Palgrave Macmillan imprint is published by the registered company Springer Nature Switzerland AG
The registered company address is: Gewerbestrasse 11, 6330 Cham, Switzerland

Preface and Acknowledgements

Preface

This book highlights the latest research in the field of sustainable aviation. In recent decades, there have been considerable improvements in aircraft efficiency and noise reduction. However, with the demand for both passenger and freight transportation expected to increase significantly in future years, the aviation sector is becoming a growing source of environmental problems and a major contributor to global warming.

Focusing on the need to address this mounting problem, this book discusses important new trends and outlines likely future developments in carbon emission reduction, carbon trading, and the impact of emerging technologies, as well as social, legal, and regulatory changes as they pertain to the aviation sector.

The book offers an invaluable reference guide for practitioners, regulators, academics, and students alike, in fields ranging from business and engineering to the social sciences. It can be used as a textbook and will benefit anyone interested in the future of aviation and our planet.

Acknowledgements

We acknowledge the financial support provided through the David O'Brien Centre for Sustainable Enterprise and the John Molson School of Business at Concordia University. In addition, we greatly appreciate the research and administrative assistance provided by Kalima Vico and

Tyler Schwartz as well as the excellent copy-editing and editorial assistance we received from Adele Dumont-Bergeron. Finally, we feel greatly indebted to Arlene Segal, Joseph Capano, Stephane Brutus, Anne-Marie Croteau, and Norma Paradis (all at Concordia University) who in various ways supported this project.

Montreal, Canada	Thomas Walker
Bari, Italy	Angela Stefania Bergantino
Montreal, Canada	Northrop Sprung-Much
Ferrara, Italy	Luisa Loiacono

Contents

Editors and Contributors

About the Editors

Thomas Walker is a Full Professor of Finance at Concordia University, Canada. He previously served as Associate Dean, Department Chair, and Director of Concordia's David O'Brien Centre for Sustainable Enterprise. Prior to his academic career, he worked for firms such as Mercedes Benz, KPMG, and Utility Consultants International. He is well published with over 50 journal articles and books.

Angela Stefania Bergantino is a Full Professor of Transport Economics at the University of Bari. She holds a Ph.D. in Transport Economics. She directs national and international projects and has been active as a consultant for private and public organizations. She has authored over 50 peer-reviewed articles on public economics, transport economics, and regulatory issues.

Northrop Sprung-Much is a research staff member at the David O'Brien Centre for Sustainable Enterprise and a Ph.D. candidate in Finance at the John Molson School of Business at Concordia University. He holds an M.B.A. (focus in Finance) from Concordia University. His research interests center around risk management, carbon trading, and environmental economics.

Luisa Loiacono holds a Master of Science in Business and Management from the University of Bari and a Master in Green Management, Energy and Corporate Social Responsibility from Bocconi University. In the past, she has collaborated with the Renewables Grid Initiative in Berlin and Accenture in Milan. Her research interests include carbon trading, environmental politics, and environmental sustainability in the aviation sector.

Contributors

Sabrina Armenio is a Ph.D. student in Applied Economics at the University of Bari. From 2016 to 2017, she was a visiting research student at the Department of Applied Economics at the Autonomous University of Barcelona (Spain). Her main research interest lie in environmental experiments focused on the transport sector.

Thomas Budd is a Lecturer in Airport Planning and Management in the Centre for Air Transport Management at Cranfield University, UK. His main areas of expertise include environmental planning and policy, airport surface access, and resilience and strategic planning. He has contributed in 11 articles and one book.

Kenneth Button is a Professor of Public Policy at the George Mason Schar School of Policy and Government and a world-renowned expert on transportation policy. He has published, or has in press, some 80 books and over 400 academic papers in the field of transport economics, transport planning, environmental analysis, and industrial organization.

Larysa Cherniak is an Associate Professor in the Department of Ecology at the National Aviation University in Kiev, Ukraine. She holds a degree in engineering with the specialty chemical technology of fuel and lubricating materials. Her thesis is entitled «Development of Technology for Petroleum Vapors Recovery Using Organic-Silicon Adsorbents».

Dimitrios J. Dimitriou is an Associate Professor in the Department of Economics, Democritus University of Thrace (DUTh). His specialization is planning, management and economics in transport, providing dedicated expertise in the air transport sector. He has contributed to more than 200 papers and presentations, and more than 20 research projects. He has also been in numerous management positions and scientific committees.

Marina Efthymiou is an Assistant Professor in Aviation Management and Course Director of the Executive M.Sc. Aviation Leadership at Dublin City University, Ireland. Prior to this, she worked for the University of West London, UK, and EUROCONTROL in Brussels. Her research interests focus on aviation governance and policy, environment, ANSPs, and air traffic management.

Selçuk Ekici works as a research assistant at the Faculty of Aeronautics and Astronautics at Anadolu University in Eskisehir, Turkey, where he is working on his Ph.D. His research areas include thermodynamic analysis, alternative/renewable fuels, and gas turbine test setups in aviation.

Kaloyan Haralampiev is an Associate Professor at the Department of Sociology of the Faculty of Philosophy at Sofia University. He holds a Ph.D. in Statistics and Demography. He is well published with over 10 publications. His scientific interests are mainly in the field of probability theory, Bayesian statistics, and demographic statistics.

Andreas B. Hardeman is an international aviation and aerospace lawyer specializing in environmental and sustainability matters. He has represented commercial aviation interests for over 20 years in IATA, ECAC, ICAO, UN FCCC, and UN COPUOS. Andreas regularly lectures at universities, speaks at industry conferences, and contributes to leading academic journals.

Arif Hepbasli is a Professor and Head of Energy Systems Engineering at Yaşar University, Turkey. He has 10 years of experience as head engineer, assistant manager, consultant, and maintenance manager. His research focuses on energy/exergy efficiency and management, utilization, and the potential of renewable energy sources and technologies. He has published over 600 papers.

Mario Intini is a Ph.D. student in Applied Economics at the Department of Economics, Management and Business Law, at the University of Bari Aldo Moro, Italy. In 2016–2017, he was a visiting researcher at the Department of Applied Economics at the Autonomous University of Barcelona. His research interests are: industrial organization, transport economics, and regulation.

T. Hikmet Karakoç is a Full-time Professor at Anadolu University. He has taken part in numerous industrial projects as a researcher, consultant, and project manager. He has published over 300 papers and 40 books.

He has a wide range of research interests, including sustainable aviation, aircraft propulsion system, renewable energy, and energy economics.

Luca Mantecchini is Assistant Professor at the Department of Civil, Chemical, Environmental, and Materials Engineering (DICAM) and teaches Traffic Engineering and Transport Technique and Organization at the University of Bologna. He is member of the board of directors of the University of Bologna. He has coauthored over 50 articles. His main research areas include traffic flow analysis and modeling, air transportation, and sustainability in transportation.

Andrea Morone received his Ph.D. in Economics at York University in 2008 and is now Full Professor at the University of Bari. His main research areas are focused on questions in the economics of decision making under risk and uncertainty, with a particular emphasis on the use of experimental methods.

İlkay Orhan is an Assistant Professor in the Faculty of Aeronautics and Astronautics. He received his B.Sc. and M.Sc. in the School of Civil Aviation and Graduate School of Sciences at Anadolu. He has published many papers and presented his work at national and international conferences. His research areas include aviation, aircraft maintenance, simulation, decision support systems, and optimization.

Walter Palmer retired from being an airline pilot in 2008. He served on the Canadian airline pilots' professional association as an executive council, as a technical and safety specialist, and as a chief pilot. Walter is a writer and speaker on issues related to climate change. He was selected in 2006 to be trained by Al Gore.

Andreas Papatheodorou is an academic researcher and industry advisor in air transport and tourism economics and management. He is currently a Professor in Industrial and Spatial Economics with Emphasis on Tourism at the University of the Aegean, Greece, and Director of the M.Sc. Programme in Strategic Management of Tourism Destinations and Hospitality Enterprises at the same university.

Marco Percoco is the Assistant Professor at the Department of Policy Analysis and Public Management at University Bocconi. He was a visiting scholar of the Regional Economics Applications Laboratory at the University of Illinois and in the Department of Geography and Environment at the London School of Economics. His research interests include transport and economic geography.

Maria Nadia Postorino is a Professor of Transport Engineering and Director of the Network and Complex Systems Laboratory at the University of Reggio Calabria. She is part of the Scientific Committee of the Ecole Nationale des Ponts et Chaussées. Her research interests include transport demand analysis, air transport systems, and computer science techniques to solve transport issues.

Margaryta Radomska is an Associate Professor at the Department of Ecology of National Aviation University. She holds a degree in Environmental Safety Engineering with her thesis «The Enhancement of Fuel Provision Objects in Environmental Safety». She has worked at the Ivano-Frankivsk National Technical Universities laboratory as an environmental engineer.

Maria F. Sartzetaki holds a Ph.D. in transport infrastructure economic development. Her professional work focuses on transport systems planning, management and economics, especially in the aviation industry. She has occupied multiple advisory positions in the transportation sector. She has contributed to more than 50 papers and conferences and more than 5 research projects with universities.

Stela Todorova holds a Ph.D. in Economics from the University of National and World Economy in Bulgaria. She is a Professor at the University of Plovdiv, Faculty of Economics. Her research interests include the management of regional development, ecology and sustainable development, and the relationships between air transport and regional development.

Kalima Vico is a research associate at the John Molson School of Business at Concordia University, Montreal. She is finishing her Bcomm. in finance with a concentration in economics. She has served in many projects within the Finance Department, helping to collect and analyze data, create graphs, and edit the write-up of over 45 research papers and projects. In addition, she has actively supported the editing team for both this book and two other edited book collections.

Nicola Volta is Lecturer in Air Transport Management at Cranfield University where he teaches in the M.Sc. and Executive courses. He is currently module leader for research methods and statistics. He has contributed to 10 articles. His main areas of expertise include applied economics, applied operations research, and productivity analysis.

List of Figures

Chapter 3

Chapter 4

Chapter 7

Chapter 8

Chapter 9

Chapter 10

Chapter 13

Chapter 14

List of Tables

Chapter 8

Chapter 10

Chapter 11

Chapter 12

Chapter 13

CHAPTER 1

Recent Developments in Sustainable Aviation: An Overview

Angela Stefania Bergantino, Kalima Vico and Thomas Walker

Climate change is arguably one of the biggest problems humankind is facing today. To address and hopefully mitigate the rise in global temperatures, we have to develop solutions that include changing our own personal habits toward more sustainable practices. In addition, on a broader scale, we have to examine and change the way certain industries are operating and how they are regulated.

Aviation is one of the main forms of transportation. Due to the increasingly global nature of commerce and tourism, there is an increased need for air travel and transport, to move both people and resources over large distances. As the aviation industry grows, the need for that growth to be environmentally and socially sustainable is

A. S. Bergantino
Department of Economics,
Management and Business Law, University of Bari, Bari, Italy
e-mail: angelastefania.bergantino@uniba.it

K. Vico · T. Walker (✉)
Department of Finance, Concordia University, Montreal, QC, Canada
e-mail: twalker@jmsb.concordia.ca

T. Walker et al. (eds.), *Sustainable Aviation*,
https://doi.org/10.1007/978-3-030-28661-3_1

imperative. The chapters we have collected in this book highlight the importance of sustainability in aviation, offer some criticism on the measures and policies used to enforce sustainability, and provide recommendations and in some cases visionary frameworks on how to best handle the sector's imminent expansion. Contributors include pilots, experts in sustainability, consultants, and academics who share the same concern for the future of our planet and who offer solutions and thoughts aimed at ensuring a sustainable future for the aviation industry.

The book opens with a leadership piece by Walter Palmer, entitled "Sustaining Flight: Comprehension, Assessment, and Certification of Sustainability in Aviation." The chapter focuses on the difficulties of defining sustainability and the urgency to clarify the term in order to create less subjective regulations. In addition, it offers a constructive criticism of industry organizations attempting to reduce the emissions footprint in the aviation industry. The author emphasizes the need for the aviation sector to achieve a zero-carbon footprint.

Kenneth Button's following chapter, "Boulding, Brundtland, Economics, and Efforts to Integrate Air Transportation Policies into Sustainable Development," adds to Palmer's ideas by highlighting the need of acting now. He offers a constructive and inspiring piece by examining the Brundtland Commission Report, discussing the notion of sustainable development and the transition to a sustainable future, and reflecting upon Pigouvian market failures and other challenges of sustainable aviation.

Following this chapter, Thomas Budd, Mario Intini, and Nicola Volta present the issue of externalities that the aviation industry is plagued with in their chapter "Environmentally Sustainable Air Transport: A Focus on Airline Productivity." They offer a discussion on how these externalities (e.g., noise pollution, CO_2 emissions, and waste generation) are an impediment to growth. They argue that reforms that intend to ensure the sustainable growth of the aviation industry need to strike a balance between the environmental and economic aspects of air transportation.

The following chapter, titled "The Analysis of the Sustainability Commitment Formulation and Implementation for the Selected Airlines," by Margaryta Radomska and Larysa Cherniak specifies the need for applying consistent environmental management within a suitable framework. They perform a detailed analysis of 60 documents issued by 35 airlines in which the respective firms discuss their approach toward sustainable management and explore the effectiveness of each airline's sustainability strategy.

In the following chapter, titled "Environmental Policies in European Aviation: A Stakeholder Management Perspective," Marina Efthymiou and Andreas Papatheodorou examine two European policies, namely the Single European Sky (SES) and the European Union Emission Trading Scheme (EU ETS). They perform a stakeholder analysis to allow for a detailed understanding of the effects of each policy and to evaluate the transaction costs within both of them.

Angela Stefania Bergantino and Luisa Loiacono dive further into the EU ETS with their chapter "Market-Based Measures: The European Union Emission Trading Scheme and the Carbon Offsetting and Reduction Scheme for International Aviation." They provide a detailed exploration of the various components of the EU ETS, review its development over time, and critically discuss the impact of the January 2019 introduction of the Carbon Offsetting and Reduction Scheme for International Aviation (CORSIA). They perform a series of experiments to demonstrate how the auction system works and how the market arrives at an equilibrium. In addition, they provide an overview of recent price trends in the EU ETS and, with the help of a linear regression, highlight several problems from which the Emission Trading System suffers.

The chapter "Impact of Air Transport on the Noise Level in the City of Sofia" by Stela Todorova and Kaloyan Haralampiev presents a case analysis of Bulgaria's capital Sofia. In their paper, the authors highlight the problem of noise pollution in Sofia and offer recommendations on how to best address this important externality. Their study uses detailed data on noise levels and air traffic which they analyze with a series of statistical models aimed at exploring their connection.

The subsequent chapter by Dimitrios J. Dimitriou and Maria F. Sartzetaki, titled "Social Dimensions of Aviation on Sustainable Development," explores the economic and non-economic challenges the aviation sector faces in its transition toward sustainability. They use a top-down approach and assessment methodology in order to evaluate the social benefits of aviation in accordance with the Sustainable Development Goals. They focus on various qualitative indicators such as human resource development, the social effect on communities, climate change, connectivity, safety, service quality, and system availability in order to evaluate the social dimensions of aviation.

The chapter "An Element-by-Element Approach for a Holistic Estimation of the Airport Carbon Footprint" by Maria Nadia Postorino and Luca Mantecchini examines airport emissions by performing an element-by-element (EbE) analysis of the various functions an airport performs. As a result, they are able to estimate just how big the carbon footprint of an airport really is.

Relatedly, in their chapter "Can Laboratory Experiments Help in Evaluating Emission Trading Schemes? A Pilot Experiment on Aviation Allowances: Lessons to Be Learned," Sabrina Armenio, Angela Stefania Bergantino, and Andrea Morone explore whether laboratory experiments can be effective in evaluating emission trading schemes such as the EU ETS. They offer recommendations on how to construct the most effective experiment and discuss the benefits of doing so.

Our last three chapters take a refreshing look at new approaches that are being implemented and positive changes that are taking place in the industry. First, Selçuk Ekici, İlkay Orhan, Hikmet Karakoç, and Arif Hepbasli discuss energy requirements and the need for alternative fuels in the aviation sector in their chapter "Milestone of Greening the Flight Path: Alternative Fuels." They argue that in the future, a given country's development will depend to a large extent on the sustainability of its energy. In their study, they focus on alternative and renewable fuels, emphasizing their differences, and comparing their advantages and disadvantages. They conclude that the criticisms surrounding some alternative fuels can likely be overcome by further technological advancements in future years.

Marco Percoco follows the search for solutions with his chapter "External Benefits of Smart Payment Tools for Airport Train Services" in which he attempts to evaluate the advantages of using advanced payment services in airport commuter trains. In his paper, he explores the benefits smart payment systems provide in the form of reduced transaction costs and reduced waiting time. He does so by evaluating a survey on the topic and by performing a cost–benefit analysis.

The final chapter in our book, entitled "Sustainable Alternative Air Transport Technologies," by Andreas Hardeman showcases some of the recent (and forthcoming) advancements in the aviation industry, including supersonic flights, hybrid planes, and cargo drones. Hardeman offers a critical discussion of these developments from both a legal perspective and in terms of the need for new regulations. He notes that the classification of these new technologies is challenging, e.g., with respect to the applicability of ICAO's Standards and Recommended Practices.

Ultimately, the chapters featured in this book should be of interest to anyone who works in or deals with the aviation industry or who conscientiously uses air transportation. The chapters not only provide ample food for thought, but also offer detailed suggestions for change by highlighting, e.g., the sustainability and economic efficiency of current industry standards and by critically evaluating the externalities of aviation. Air transportation is clearly expanding, as is its potential impact on the environment—which is why this topic is more important now than it was ever before. As a passenger, you can evaluate how sustainable a given way of traveling is and make an informed decision about which mode of transportation (e.g., rail vs. plane) or which carrier you want to use. In addition, as a regulator/policy maker/aviation professional, you can change the industry from the inside. We hope that our book provides readers with the inspiration to do so.

CHAPTER 2

Sustaining Flight: Comprehension, Assessment, and Certification of Sustainability in Aviation

Walter Palmer

Global warming and climate change preoccupy us and figure very prominently when we think about sustainability. That is especially true when considering the energy-intensive nature of commercial aviation. Many commercial air carriers, governments, regulators, and industry organizations therefore engage vigorously on the matter, striving for reduced fuel consumption, and also new fuels that are lower in net carbon content. At the larger international scale, when we examine the activity of the International Civil Aviation Organization (ICAO), we see that the focus has been CORSIA—the Carbon Offsetting and Reduction Scheme for International Aviation. This policy initiative is not in full effect at this point; therefore, it is difficult to assess it properly, but the rationale for CORSIA was excellent: The industry needed sustainable flight energy—fuel—but while that new fuel technology developed, and to satisfy emissions reductions goals that commercial aviation wanted to get started on quickly—no emissions increase from 2020, 50% reduction in emissions by 2050 (IATA, n.d.)—some sort of offsetting or carbon

W. Palmer (✉)
Independent Writer and Researcher, Guelph, ON, Canada

T. Walker et al. (eds.), *Sustainable Aviation*,
https://doi.org/10.1007/978-3-030-28661-3_2

reduction crediting would be needed (IATA, n.d.). And markets for carbon credit instruments already existed as examples. But, it is possible that preoccupation with offsetting has had the effect of leaving the fuel matter somewhat to one side.

Sustainable fuel ventures and carbon offset schemes are, in some ways, difficult to compare. The former produces, for the air travel industry, an actual amount of the fuel commodity. It does this in a way that reduces net carbon emissions. By contrast, the latter creates an economic circumstance for the introduction of a measure that prevents carbon emissions or causes the absorption of atmospheric carbon. But offsets take place independent of geography and have nothing directly to do with flight. There might be a hydroelectric project that promises to reduce a country's power generation emissions. Or maybe there is a project to plant trees. But an essential point, here, is that while fuel and offsets have certain differences, the production of fuel and the activity that an offset is designed to fund are activities that take place in the physical and human world and are (or should be) assessable on a common basis in terms of sustainability.

If standards for offsetting initiatives—including matters related to the sustainability of the actual projects that are funded—are not rigorous, they perhaps allow a certain laxity that might favor different jurisdictions or organizations that want their particular scheme to be included where a more demanding standard and certification process might leave them aside. This has the effect of reducing overall progress toward sustainability. Compounding that evolution, there may be a desire to apply such sustainability standards to alternative fuel ventures, too (Lyle 2018). And it is true that a virtually identical sustainability standard must apply to things that substitute for one another in aviation. Otherwise, good fuel programs could be gamed by lax offset programs, or vice versa.

Considering the capital requirement and the time it will take for new, non-kerosene-burning air transport to be developed (Palmer 2016), reduced-carbon fuel is the only answer we will have during the time frame within which carbon mitigation is generally considered: to 2050. If massive, cheap, technical atmospheric carbon extraction becomes feasible quickly, the equations change. There is no apparent evidence of any such significant or imminent capacity. But carbon can be extracted from the atmosphere at great scale through such things as land use modification that results in banking quanta of carbon. The resulting commercial value of such carbon sequestration and other carbon-reducing activities, as offsets, through policies, such as the Clean Development Mechanism (CDM), has been low (Martin et al. 2016). Nevertheless, it should be kept in mind that in a notional configuration of climate change mitigation efforts where actual emissions had become low, sequestration would retain its value if there was a general effort to render the global economy 'carbon-negative'. In that regard, it should be noted that methods for reducing atmospheric carbon are valuable in their own right—not simply for offsetting. Logic dictates that we must accept the urgent need for the development of sustainably produced low-carbon fuel, so that we can take advantage of everything—low-carbon fuel and other ways of drawing down atmospheric carbon. Right now, demand for all the available carbon credits is low, and obviously, the people, governments, and organizations that produce them would like to see that change. So, it stands to reason that aviation is an obvious potential customer because airlines will need offset credits for reducing the effect of their emissions. It is that admitted short-term industry need for carbon crediting that justified CORSIA. But, ultimately, achieving the goal of low/zero-carbon fuel is the answer. To the extent that an industry (or the whole economy, for that matter) depends on offsets, the opportunity to have other means of carbon reduction or sequestration produce a better—perhaps even carbon-negative—result is lost. A discussion on how to ensure that carbon offsetting initiatives have commercial value is needed, but that shouldn't detract from the desirability of actual low-carbon flight. And while it is generally acknowledged that sustainable aviation fuel (SAF) is important, it may be that offset discussions have preoccupied ICAO talks, while the use of SAF now seems to be regarded from the perspective of, and comparison to, offsetting (Lyle 2018).

So, that is a summary of two potential problems with offsetting: First, that offsetting has a troubled history in terms of credibility (Becken and Mckey 2017) so, a reliable, first-rate system for certifying offsets' true merit is absolutely needed. Second, that some initiatives toward sustainable fuel, while advancing quite well and focused on real, authoritative, broadly subscribed sustainability standards and certification, are not getting, perhaps, as much support as they deserve.

Before going further, let's mention that Emissions Unit Eligibility Criteria (EUCs) have been agreed at ICAO and promulgated in March 2019 (ICAO 2019). These are general guidelines about what conditions would need to be met by a qualifying offset program. But the EUCs do not incorporate detailed sustainability requirements.

We should remember that CORSIA discussions at ICAO focus upon the concerns of national delegations and not necessarily, in each detail, upon aviation or the larger picture. That's understandable, each country comes to any set of talks with many things in mind. For some delegates, looming large is the fact (just alluded to) that offsetting already exists but is not generating desired levels of revenue for them. This is an entirely valid worry, but it has a tendency to inform the discussion wholly. It would be natural for some governments in states where offsets are generated to want aviation to be offset-dependent. This unfortunate combination of focus by certain member states on offsetting, including their wish to legitimize and market existing offset schemes that are perhaps less credible, is definitely a concern. Second (and this flows from the first), since extant offsetting schemes might operate to less stringent standards than those being considered by some members of the aviation community, and since stakeholders want their schemes to remain valid, they would likely desire offsetting standards that are their own. If standards are not promulgated universally (just like ICAO's current EUCs), we cannot be sure that they will be strong and reliable. The discussions seem to have fallen prey to a lower common denominator tendency—some parties not even favouring what is referred to as a 'vintage' provision that would shut out some older offset schemes that have been determined to be weak. In that regard, specifics are yet to be decided and may be decided in an appropriate way but, generally, if individual countries have any success in advocating for only the most general sustainability guidelines and, within such general EUCs, having their own authority to develop their own particular detailed standards and methodology for applying them, nothing of great value can be achieved by offsetting because the good will be mixed in with the less good. Then, if fuel projects are also assessed in the same way, the damage is complete.

Parenthetically, as politically difficult as it might be to achieve, a proper sustainability standard, standing beside or within the ICAO CORSIA EUCs, would end discussion about whether offset schemes needed to meet such things as vintage provisions.

That is a brief, broad-strokes assessment of a complex process that has taken years to evolve. In many ways, it constitutes a mere impression because the 'facts' are so vulnerable to subjective assessment, and—very importantly—so much of ICAO discussion of these important matters has taken place behind doors that are closed to interested parties (Laville 2019). But when all the paras and sub-paras and each 'whereas' and the blend of umbrage and good will has been sifted, there is a great deal of disappointment. The people who are genuinely interested in making commercial aviation sustainable are not all entirely convinced that the CORSIA policy efforts have succeeded. The work done by ICAO itself, together with many delegate members, has been monumental and commendable. It is difficult to bring disparate voices together, and no fault can be assigned if the result is not really suitable. And grave doubts do exist at this writing (Greenair Online 2019).

If ICAO talks have been rough and less productive than one would wish, the work done by many actual industry-involved players—airlines, fuel companies, specific aviation-focused governments—has been quite remarkable. But unless larger (presumably ICAO) consensus is formed around the SAF project with a real, rigorous, and complete sustainability standard, and one that is carried over to offsets in a lockstep way, sustainable aviation will only be achieved by establishing and then expanding a core group commitment to SAF and sustainability itself. Commercial aviation is approximately one hundred years old. An industry that has compelled itself to face the greatest technical and human challenges imaginable with regard to safety and performance—and has by-and-large succeeded magnificently—will presumably prevail, whether through ICAO policy generation or otherwise. If there are popular, societal ideas about GHG emitters and their respective obligations, commercial aviation's role in global warming and climate change will suffer a progressively higher profile, so there is urgency from the point of view of the industry's commercial prospects.

Now, beyond flying's plight is the unforgiving nature of the planet's and global and national societies' emergency. The amount of time and resource that we can saw off in order to make the climate challenge easier for one particular country or another, or for this or that industry or company to address is exactly zero. In fact, it is less than zero; we are well into the dynamic of rising global temperature. What we are attempting is to minimize the harm and prevent wholesale ravaging of the ecosystem and human society. But while the problem can't be 'fixed,' it is very important to ensure that the planet is as habitable as possible for as much of humanity and non-human life as feasible. In a perfect world, we might have understood threatening climate vagaries more than a hundred years ago, and we would have known exactly what to do in order to prevent industrialization from becoming such a destructive force. That boat may have sailed but the work that we can do is extremely important.

The Intergovernmental Panel on Climate Change (IPCC) Special Report of October 2018 (IPCC SR-15) outlines the gravity of increasing global warming in the context of a 1.5 °C global atmospheric temperature rise, as opposed to the 2 °C rise that policy had previously contemplated. The thrust of that report is that we have to get net greenhouse gas (GHG) emissions reduced by 45% by 2030—eleven years, at this writing—and down to zero by 2050 (Intergovernmental Panel on Climate Change 2018). Even among those who have supported the effort to address global warming vigorously, this is shocking.

Frustratingly, as great as the climate challenge is, it fits integrally into something that is larger, and our efforts cannot succeed if we fail to grasp that bigger picture. This is about sustainability, and how it can be reliably achieved. Fuel is our entry point; production and use of energy for flight constitutes the greatest single stone within the wall that stands between commercial aviation and its quest to become a more benign presence in the world. Shattering that stone will open a hole in the wall. But it will also allow for the tumbling down of the whole edifice if we manage to break that stone in a way that illuminates the path to the similar removal of all the others. We cannot make low-carbon fuel a part of the sustainable aviation project unless we also ensure that the fuel comes in a way that is, itself, sustainable. The good news is that if we proceed on those lines, everything that we do serves as a model for every other. If we launch off without considering that, everything that we do will make a bigger mess. Because, as important as fuel is, every other activity that goes into the delivery of commercial flight and—in fact, every activity in every industry—must meet the sustainability requirement, eventually.

So, perhaps all this will help us understand the urgency of having a uniform, consistent approach to the broader idea: sustainability—that property of an action that determines that it can be both continued and supported.

Of course, we focus on GHG reduction, but that can result in a narrow vision that (in the longer run) makes all legitimate goals (including GHG reduction itself) harder to achieve. Preoccupation with the major task of GHG reduction to the exclusion of other things can lead to the neglect of essential matters; if we reduce emissions by doing things that are not, in themselves, and in other respects, sustainable, we can cause compounding problems. For example, if we engage heavily with a technology and an attendant group of suppliers whose work can never get carbon emissions reductions greater than 50% and then also realize that their processes harm the environment in other ways while also disrupting society, we will have built an economic and political trap; it is expensive to walk away from the production capacities that the suppliers have created, and it is politically hazardous to abandon the constituency of workers that has grown with such an industry.

If we consider where sustainability efforts are right now, we discover that we are manifestly better at recognizing certain particular qualities of sustainability than understanding what it takes to ensure that those qualities materialize together. We can recognize a good quality in a new procedure or technology, but we are not good at seeing if every aspect of the new thing is good. What filter can we propose for human activity so that the actions that are not sustainable in the most comprehensive sense are caught? What tests can we apply? What attributes can we force that necessarily result in action that is perpetuable and supportable?

To talk in terms of 'Sustainability,' the first barrier is the inconsistency in how that term is regarded and the arbitrariness of our manner of pursuing it. So, it will be argued here that sustainability should be sought only in broad context and comprehensively. In coarse terms, we know that continuing to put greenhouse gas (GHG) in the atmosphere creates such an enormous effect that it cannot continue, and that efforts to maintain the existing fossil fuel energy regime cannot be supported. But if the sustainable production of low-carbon flight energy is simply the first thing that we think of, it also serves that illustrative purpose, showing why, for example, environmental difficulties cannot be resolved without social justice.

Many people who have sustainability in their brief likely feel that they know sustainability. But what does it really mean when someone says that their company or product is 'sustainable'? What about the related claims by governments in the area of policy? We don't always know exactly what they mean because we don't know how they understand the word 'sustainable'. Nor do we know how faithfully they are putting even a flawed view into action. Since they don't all say the same things about sustainability, there cannot be a common understanding. This is a serious concern because we all know that this lack of rigor affects everything in our consumption-driven world—it fosters skepticism and cynicism and militates against the traction that the concept of sustainability should enjoy in steering rational consumer choice. Also, to some extent, it establishes a feedback loop at levels of decision beyond the consumer: It tells policy makers and corporations that sustainability initiatives can never achieve full value perception—that if sustainability costs money or effort, it may not be worth it.

What would help the effort in having everyone understand and accept sustainability in a more congruent way? If there are many attributes to sustainability, and if they interact, my view is that it is more helpful to recognize sustainability in terms of its operating dynamic rather than as an abstract idea with our focus simply upon the distinct, isolable characteristics that often preoccupy us. The constellation of questions that many of us have about sustainability then shifts from parsing its appearance from one aspect or point of view and moves toward figuring out how it works. Merely, deciding upon sustainability attributes provides an incomplete evaluation: We need to know why that attribute helps an action to be considered perpetuable and what other attributes must then go along with it. Once we better understand how sustainability works, as a multi-dimensional quality of action, we can filter and test for sustainability more effectively.

To increase the awareness of the importance of having a comprehensive idea of sustainability, let's invoke practicality for a moment and go back to GHG emissions: Saying that the world must achieve net zero emissions in a very short time is so startling that it sounds more like environmentalist extremism and advocacy than a 'realistic' legitimate policy goal. Can it be realistic to suggest that we get emissions to zero so quickly?

Check that word, 'realistic'. SR-15 shows us that the goal responds to reality as opposed to what is realistic. Letting the globe's atmosphere run to more catastrophic high levels, where compounding effects can render its horrific evolution completely unbound, is just unacceptable. In fact, the report says that we need net negative emissions. That is the hard reality that is harder than all of the economic and financial and political realities that are presented in counterpoint. Commercial aviation's commitments to stop increasing total emissions by 2020 and to achieve an absolute reduction of 50% by 2050 seemed laudatory and even ambitious a mere few years ago; now, SR-15 shows them to be perhaps slightly modest. The great efforts that we have made in establishing agreed policies and goals are insufficient. Former UN Secretary General, Ban Ki-moon, spoke about his '50-50-50 Challenge': a 50% reduction in carbon emissions by 2050, in the face of a 50% increase in world population (Ki-moon 2010). Now, we must consider that this was not good enough, and that we have been far behind in our thinking and achievements. That failure is fear-inducing in almost everyone who understands it, but we can see that the approximate 1 °C atmospheric temperature rise that has already occurred is now producing effects that we would not have considered 'sustainable' if we had recognized, long ago, that they would occur over time (Union of Concerned Scientists, n.d.).

The point is that the global economy that is producing this massive threat is immensely complex and we don't really see an easy way of separating carbon production from the rest of what we do. And, if absorbing these facts swamps our thinking, it is important to accept the message but keep our heads above it. It may be disorienting, but we do not really have time for denial, anger, bargaining, and depression.

To avoid missteps, action depends upon clear and broadly endorsed terms for understanding sustainability in its largest conceptualization, together with the development of a way to assess and certify activities in those same terms.

Faced with an activity that seems unsustainable, the only questions are: (1) Do we continue the activity as a very long-term proposition (helplessly, and as long as it can be perpetuated) and suffer, indefinitely, those negative outcomes that logically dictate that we should have stopped it? Or (2) do we continue the activity in the belief that the world we will have thereby changed is the one we are 'supposed' to have? Or (3) do we change or stop the activity because we believe that the outcomes are unacceptable? Generally, the rational response is to say yes to 3. But, generally, we fail. Why do we fail to choose correctly?

Much of the answer lies in short-term selfishness and greed, but also in that inability to perfectly understand and describe sustainability. To a great degree, we fail because of imperfect knowledge about the challenge, not comprehending the problematic consequences of continuing something that is—if it were properly considered and understood—unsustainable, or we fail because we do not understand how to change: to make the activity sustainable or to substitute another activity.

As to urgency, if we recognize that we must stop or change the activity, we know that we must get started immediately, and yet proceed on the basis of imperfect knowledge: get going, but make no mistakes. The point is to accept that this increases the burden of care that we exercise so that our efforts in the pursuit of sustainability produce as few missteps as humanly possible when we are acting on the basis of the best knowledge available and incorporating new knowledge, as it appears, in our definitions and standards. If we proceed some distance in a direction that does not ultimately serve the goal pursued, we can create further, new, unforeseen problems that must then also be solved. So, we must bring caution strongly into play, right in the face of urgency.

How do we characterize aviation, or its energy input, in terms that are sufficient to allow us to get started on the problem without delay? If we understand some of what sustainability actually means (first step), we can then recognize good examples of practice (second step). In other words, we can start by postulating things about sustainability and examining models in order to elaborate our understanding; prominent initiatives have likely consolidated much of the knowledge that we need about sustainability and what we need in order to accomplish it.

Are many sustainability frameworks badly flawed? Sure. We expect that existing sustainability framework efforts have faced the 'real vs realistic' question discussed earlier; they define sustainability and parameterize it based upon subjectively formed ideas about what is 'practical,' 'affordable,' or otherwise 'feasible'. In too many cases, we see that our definition of sustainability is really a demand of it. It is as if we ask an airplane to carry loads and attain speeds according to our wants, expectations, and the hard realities of our balance sheets—rather than the even harder realities of physics. Our understanding of sustainability can't be allowed to become a mere recital of popular notions and convenient approaches. It must come from pursuing implications of actions as far and deeply as we can.

It is true that we often accept that standards are generally based upon need. Now, if need is measured relative to the difficulty or expense of achieving it, the standard is an accommodation and there is often nothing wrong with that. Standards, even for extremely important things, are often set this way. Standards for aircraft certification and pilot proficiency—even though both are critical—are set according to what is achievable because perfection is unachievable and we have to decide what we can reasonably accomplish. But as the criticality of a concern becomes higher and higher, our standard setting also becomes more critical and demanding. As our level of need for high performance or standard approaches the absolute, the term 'reasonably achievable' starts to lose meaning. The greatest need imaginable implies the greatest commitment, effort, performance, and standard imaginable. For commercial aviation or any government or industry to set a sustainability standard for activities that affect the very viability of the Earth's prevailing environment, based upon rough and differing assessment of what is 'reasonably achievable,' is not rational. If we have at least a 'nose' for sustainability, we can use even poor standards to help us develop by being attuned to weaknesses.

Before we think about models, let's think about the word: 'sustainable'. In terms of the longest established meanings for it (setting aside for a moment anyone's take on a current topical meaning, and whether we speak of single discrete occurrences or ongoing activities or processes), there are two general strands of connotation: (1) physical possibility and (2) moral aptness. We can assess or measure a human activity as something for which the requisite physical resources in material, energy, power, and strength are available, and we can also assess and measure the activity as to whether it is moral, or fair and proper to perform it. These assessments are guided by our understanding of effects: Is it possible for me to continue this effort? What happens to people and the environment if I do? Or, Can I do this, and should I? From the point of view of those acted upon: Can a certain person or institution or entity bear this, and should they be required to? (OED 1971).

That is an extremely intricate set of considerations. As well, and in purely dictionary terms, we generally consider the moral dimension as distinct from the physical one; sustainable as morally valid has a different meaning from the concepts of physical supportability, endurability, and perpetuability. Yet, it is true that we can look at any action from either a moral or physical possibility standpoint.

What is perhaps new for some people—even some of those with professional practice in sustainability—is that we must use both dimensions (physical capacity and moral suitability) for the generation of policy. The questions must be both, 'Is this action, or activity, or policy, socially just?' along with, 'Is this action, activity, or policy benign from an environmental perspective?' Why is it that we have come to invoke both meanings of the word sustainable in policy deliberations? In the time of the World Commission on Environment and Development (WCED or 'Brundtland Commission'), the focus was on the environment. But the reason that it was the driving concern was that it was a constraining factor in the face of an assumption that development was an essential social and geopolitical good; environment was the focus because development was the driver (Brundtland et al. 1987).

Recognition of development as a social justice goal and an economic ideal can leave us regarding environmental benignity as an oppositional constraint. However, we have learned (but often forget) that our environment is a dominant wealth enabler. Our economy exists within, and pulls material and energy resource from, the physical world. We may have the fear that recognition of the need for environmental benignity works against the social good of development because it forces us to consider any action carefully rather than acting immediately. In this way, environmental concerns are seen to act as a brake on development action. That is a misplaced fear, and we should think about why that is.

Jim MacNeill served as rapporteur for the WCED, and it was he who drafted the Brundtland report. He was of the firm conviction that recourse to the simplistic, so-called Brundtland Definition of sustainable development, with its focus on (1) 'needs' and (2) intergenerational considerations, tended to deprive the discussion of a more valid picture of what 'sustainability' really meant. He remarked also on how whole industries, for example, had allowed themselves to define sustainability goals narrowly and then claim enormous progress based upon achieving anything that seemed to relate to some very vague, amorphous, limited conceptualization (Palmer 2016). The thrust of the WCED report was to recognize that social development, which it acknowledged the need for, could not continue while ignoring environmental constraints, partly because we were relying on the environment for what it offered in economic terms.

The WCED certainly did see the social component of sustainability, too. However, for many people, even if they accept the social justice component as being a fact of life in the sustainability world, there has been a failure to see another logical conclusion. When they examine that report, they see little concerning the relationship between the social and environmental. They fail to see that what they otherwise regard as the independent moral and physical meanings of 'sustainable' actually operate as joint necessities. They cannot operate separately. When we talk about Big 'S' Sustainability, à la 'Sustainable Development,' sustainability of a corporation's activity, sustainability from the point of view of domestic and international public policy, we have to see that in the long term, and in the fullest conception of what we are after here, it is not really possible to ensure that an activity will be environmentally benign unless we have also ensured that it is supported as being fair—and vice versa. We cannot have comprehensive environmental benignity without comprehensive social support for the value of that goal and the efforts to achieve it. And we cannot have fair and just human society without the ideal environmental resources required to support it—the two aspects are not different; they are bound. This is an important consideration for anyone involved in any way with the questions of sustainability writ large. It's perhaps easier to understand this point if we think about all the situations in the world where people would feel that they would be deprived of wealth and advantage were they forced to protect the environment in their main economic activity.

All of this is particularly important as regards commercial aviation as its services are a fundamental part of development. An examination of the seventeen Sustainable Development Goals (SDGs) (Nam 2015) forces one to think about how many goals could be realized without commercial flight and, on the other hand, how many of the goals clearly imply a need for aviation (along with other activities) to improve its environmental viability. Regarding the former, for example, can we have more or better 'Industry, Innovation, and Infrastructure,' (Goal 9) without moving, frequently and quickly, the people who can help achieve that? And then, also, can we think about the need for 'Affordable and Clean Energy' (Goal 7) and 'Climate Action' (Goal 13) and not imagine that aviation should be pursuing these? Air transportation must be sustainable. But, also, global sustainability, in the form of the SDGs, cannot be achieved without air travel.

We can lay out a calculus that incorporates environment, social justice, and economic worth under the rubric of 'sustainability'. It may make the matter somewhat less opaque to many sustainability practitioners, theorists, and policy makers. The logical sequence of considerations would be:

> All humanity wants wealth and the benefits that accrue from development.→ Development and 'developed world' activities cannot continue if environmental damage erodes the net benefit of development *and* the resource base that enables it → Therefore, environmental benignity is essential → but environmental benignity can only be achieved with the support of the individuals, corporations, and governments that use the physical environment to meet their wants and needs → Social justice is essential to secure the support of all such human players → Wealth is an enabling resource for social justice goals → Development brings wealth → All humanity wants wealth and the benefits that accrue from development ... → etc.

The fact that this logic does not penetrate perfectly to the level of individual motivation, or industry or government policy generation, is patently unacceptable; the world needs to understand what renders a human activity sustainable and needs to execute each policy and action on the basis of that understanding. Sustainability is a concept that integrates environmental and social just factors. If it is not understood that way, it becomes difficult to achieve advance of sustainability in one aspect of our human, physical, and economic world without making some other important thing impossible or more difficult.

As to commercial aviation, this understanding is particularly vital. To see why, consider, for a moment, commerce more generally: The higher the level of production complexity, the larger the scope of upstream input and downstream output considerations and, therefore, the more difficult it becomes to assure criteria of sustainability through the production chain, down to the end product. Commercial aviation relies upon the sourcing and use of a myriad of goods and services, and each airline comprises many highly complex service production units in different locations. Furthermore, there are many airlines, and the complexity of the production chains of all these operating entities is compounded by the fact that their chains must stretch across those many geographic regions and jurisdictions. If 'Airline X,' based in 'Country Ex' operates on the common international route flown also by 'Airline Z,' based in 'Country Zed,' each airline must have a supply chain in both countries (X in Ex; X in Zed; Z in Ex; Z in Zed). If both airlines aspire to sustainability, their respective pursuits would be much easier if their vision of sustainability were shared. The challenge is to have consistency through the entire chain of goods and services sourced by each airline. By extrapolation, since each of these airlines shares other operating locations with other airlines, the challenge broadens to spreading that understanding to all airlines in each and every place from which any of them may source such goods and services. Aviation must, therefore, lead others when it comes to agreeing detailed understanding and commitment in (1) defining the characteristics that would render it an activity that can be ongoing without harm to local and global environmental and social systems, (2) establishing criteria that allow measurement and assessment of such characteristics, and (3) developing a unified standard for the application of these measurements and assessments. Such a standard could be applied to all aviation inputs, not just fuel. Commercial aviation must be one of the industries most aggressive in ensuring comprehensive, rigorous, and broadly applied standards of sustainability. Locally developed standards, notwithstanding that the guidelines that inform such standards may be generally good, cannot satisfy such requirements.

One way of regarding this would be to say that if aviation, as an industry, is not confident that it is applying the most exacting, globe-leading techniques in advancing those aspects toward the highest, ultimate goals when examining the most prominent aspects of its sustainability profile, it constitutes an industry-ruining failure. That is because (for example) if one supposedly sustainable carbon offsetting scheme or fuel project is found to be inadequate, the traveling public will feel entitled to reject every claim.

If the most important single element of aviation's sustainability profile is flight energy (fuel) and the GHG emission associated, let us examine where we are with that. Activities are being undertaken now to make flight more sustainable from the point of view of GHG production. Some initiatives reduce the need for fuel and some reduce the net carbon content of that fuel. Taken together, they are:

1. Improvement of the aerodynamic and propulsive efficiency of aircraft and engines so that less flight energy is required;
2. Improvement of the efficiency of use of airspace;
3. Improvement of the efficiency of flight planning and pilot operation of aircraft;
4. Development of new means of producing flight energy for current types of aircraft;
5. Introduction of offsetting for the emissions effects of flight energy consumption with activities that, on a life-cycle basis compensate for or reverse such effects;
6. Development of new modes of travel or new types of aircraft that reduce or eliminate the need for current fossil fuel means of producing flight energy.

This last one (6) cannot assist us with the challenge of providing sustainable flight energy during the life of both the existing fleet and new aircraft that still exploit power plants burning kerosene-type hydrocarbons. New, revolutionary flight energy regimes will surely come, but kerosene-burning airplanes will continue to be built for many years, and each one is an enormous investment so, once built, they will all fly for decades—likely far beyond 2050 (Palmer 2016).

Efficiency improvements (1–3) cannot keep pace with projected industry growth (Lyle 2018; Palmer 2016).

If the larger goal is to render air travel sustainable, anything that is undertaken toward that end must, itself, be sustainable. In that regard, both new fuel production schemes (4) and emissions offsetting schemes (5) must be sustainable—and (to the points raised in opening paragraphs) they must both be sustainable in uniform terms; if there is to be a sustainability standard for the production of alternative, low-carbon sources of fuel, any offset that is employed to obviate the need for such fuel must meet a uniformly comparable standard. This should be regarded as a fundamental principle. It should not be allowed that either offsetting schemes or new fuel technologies are able to 'game' each other in carbon accounting or broader measures of sustainability factors. As to how such a sustainability standard must be agreed, it must address all of the sustainability concerns that we can identify (highest common denominator), and it cannot be driven by the desires, preferences, or perceived 'needs' of those who would compromise in creating something less (lowest common denominator).

In summary, the commercial aviation industry must:

1. Have a clear, agreed understanding of the most comprehensive meaning of sustainability;
2. Have a sustainability standard that meets that understanding;
3. Apply that understanding and standard in the employment of a certification mechanism—whether developed for purpose or through subscription to an existing standard;
4. Assure that the standard and certification mechanism applies to all aviation 'inputs,' together with whatever substitutes for such inputs.

Does such a sustainability standard exist? No. It is not possible to say that there is a formal extant standard that perfectly comprehends every environmental and social facet of sustainability. It is argued here that two things (among others) are required to make the journey toward such a goal as efficient as possible are (1) knowledge of the destination and (2) conceptualization of the most useful first steps.

However, since we speak of examples of extant models of sustainability assessment and certification as starting places, we can certainly proceed on that basis.

Relying on our previously mentioned 'nose' for sustainability, one example of a standard that is informed and directed by a vision of what perfect sustainability seems to be, and strives ambitiously to come as close as possible to that ideal, can be advanced: the Roundtable on Sustainable Biomaterials (RSB). Much is known about the RSB, and more than can possibly be explained here should be known about it and can be easily found (rsb.org). It is probably illustrative to consider, at least, the structure of the standard and the sorts of things that it addresses. Firstly, it is many standards, varying depending upon the product being considered and the production role of the individual or organization accessing it. But in general terms, there are twelve principles, and each principle has related criteria. In turn, each criterion comes with its own set of compliance indicators. To provide an idea of the general scope of the standard, these are matters to which the twelve principles pertain:

1. Legality;
2. Planning, monitoring, and continuous improvement;
3. Greenhouse gas emissions;
4. Human and labor rights;
5. Rural and social development;
6. Local food security;
7. Conservation;
8. Soil;
9. Water;
10. Air;
11. Use of technology, inputs, and management of waste;
12. Land rights.

Incorporated within the principles, criteria, and requirements are many matters that are important in their own right. These are examples only: provisions for free, prior, and informed consent (FPIC) related to direct involvement of affected people in decisions concerning a project; land use change (LUC) and indirect land use change (ILUC); biodiversity; and genetically modified organisms (GMO) (RSB, n.d.).

That is a view of a good starting place and an excellent example in terms of sustainability assessment and certification. In fact, acknowledging that our need to get started is urgent, beyond serving as a model, the RSB standard seems ready for adoption. Key air industry, fuel industry, and environment and sustainability NGO leaders support RSB adoption for SAF. As a starting place, it will need to be adapted as our understanding of the issues improves, but it is ready to go.

Domestic aviation GHG emissions are considered part of the overall carbon budget for respective states (Palmer 2016), so individual governments may make a case that a sustainability standard for fuel used in domestic operations can be created in a particular state-driven way. By contrast, international aviation, which has committed itself to carbon-neutral growth from 2020, and real reductions to 50% of 2005 emissions by 2050, needs a broadly supported robust standard, like the RSB. Since SR-15 states that all emissions must be reduced by 45% from 2010 by 2030 and net zero by 2050, the uniform standard for fuel and offsets is needed immediately. A lax standard is simply one that does not work. With any lack of thoroughness, GHG emissions reductions targets can be missed by large factors.

Perhaps all that has been outlined here makes it clearer (if that were necessary) that initial action in establishing sustainability standards must align quite well with the ultimate goal conceived—perfectly sustainable flight energy. Investment in fuel production and attendant infrastructure can be expected to be amortized over the multi-decadal periods that have been referenced, so wrong steps will be immensely costly. They can create political barriers to realizing necessary goals, as well: People make career and investment decisions based on what is undertaken and they usually vote to protect all that. It is the ultimate goals—though most difficult to achieve—which are the easiest to identify. We know that human activity must eventually be rendered environmentally benign in virtually all respects. We know that such comprehensive alteration of our mode of existence here will require the compliance of almost every human being and every social, political, and corporate entity that humans create. We can deduce that such support will demand that commerce and politics evolve toward greater knowledge, justice, and fairness. These should be the sorts of ultimate goals that we seek in pursuing sustainability, generally, because it yields unambiguous guidance.

Whether we speak about fuel or compensating offsets, anything but an extremely high standard allows arguments about equivalency, and these can be tailored and rationalized in a subjective way. Essentially, we have to seek and pursue perfection. Perfection—regardless of how hard it is to get close to it—is not subject to approximation. In that regard, concerns are being expressed by those who watch the policy process closely.

See the letter sent by the International Council on Sustainable Aviation (ICSA) in advance of ICAO final decisions on offsets (ICSA 2019).

Current deliberation at ICAO and in many national discussions seems to focus on achieving an offset (and consequently fuel) sustainability standard through compromise. That is politically appropriate in a political body. However, the advance of sustainability from a philosophical and moral proposition to a recognition of physical realities and an attendant understanding of commercial realities in the longer term has occurred and is growing within the airlines and air industry constituency, including many of the national and local governments that serve as the corporate homes of such airlines. If broad deliberations are not producing consensus on the idea of equivalent and highest sustainability standard for both fuels and offsets, it will be left to international airlines, airports, municipalities, regions, and states that do 'get it' to come together on a standard that is actually suitable and to build alliances and support on that basis. In the larger, diplomatic sense of 'political,' it is not at all helpful to ignore ICAO processes. Multilateralism has produced some huge successes for commercial aviation, but the industry and some governments will, perforce, move past dependence upon ICAO talks for solutions if those deliberations do not produce what the industry knows that it needs.

In the context of all this, what should happen now? The RSB standard is, arguably, the most rigorous, highly developed standard available, and it is a standard that is gaining great currency and credibility with first actors in the air industry. While the RSB standard must progress from being a fixed-rate-of-reduction standard (currently, for fuels, a fixed 60% reduction in carbon emissions as compared to fossil fuels) (RSB, n.d.), to a 'performance-based' standard, that change could happen. The RSB can become a standard that provides certified information on how much carbon reduction is actually achieved by a given product. A fuel product that achieves an 80% reduction in emissions relative to a benchmark fossil fuel source can be recognized in that way ('RSB C-80,' notionally). We cannot progress toward zero on the basis of knowledge about 50% or 60%. If a zero-carbon fuel enjoys no market advantage over a 'C-60' fuel because they are both certified 'sustainable,' certification becomes less meaningful; a fuel that has very low-carbon content relative to another—say, an 80% reduction as opposed to 60%—is much better for the environment and certification should recognize that. While fuel manufacturers who see no prospect of their particular technology and products ever getting to very low-carbon ratings may object, a certification system needs to provide knowledge about what is actually being achieved by different players. In fact, since some flight energy sources upon which we must depend may not be able to reach values near zero, others must perhaps be rated to below zero carbon values so that the system, in aggregate, can meet its goals. So, a reasonable way forward is for the international air carrier community to commit to sustainable fuel manufactured to the RSB standard, for the RSB standard to make paced progress toward complete sustainability on all indices—including actual performance data on carbon content, and for the core constituency of airlines, governments, and fuel producers to advance that as the standard that airlines use everywhere they buy fuel. So, further, it would be well, if politically feasible, for CORSIA and like schemes to adopt the RSB standard as well, for both fuel initiatives and offsets.

In outlining a quest for perfection—and even in arguing for the use of perfection as guidepost—it is possible that what has been achieved is understood to have been presented in a negative light. That is not the intent here. For example, the work done by the ICAO bureaucracy, the national delegations that are its constituency, and the interested party organizations that advocate there, representing airlines, and environmental and other organizations is—as has been said—laudable. An examination of the CORSIA Emissions Units Eligibility Criteria shows the multilateral community coming together on something that is extremely meaningful. We only say that more will be required.

All of this—seemingly far too ambitious—is not pie-in-the-sky; it merely avoids a 'lie-in-the-sky'. The 'user community' of the sustainability standard that we discuss is quite clearly, and most obviously, made up of the airlines and the people who sell them fuel. The person who is responsible for deciding whether the airline industry, with its fuel partners, is behaving responsibly is the airline passenger, or the individual who challenges—on the basis of lack of sustainability—the right of the airline to operate and that passenger to travel. The world society's collective awareness concerning these matters grows inexorably.

As other industrial sectors and other transportation modes de-carbonize, and as flying increases, the adverse sustainability profile of the activity will grow. Every moment of delay means that the ultimate effort, when it comes, will be more difficult and costly. The best way forward is the only way forward: be seized of the challenge and advance in only the best way that we see.

References

Becken, S., & Mackey, B. (2017). What role for offsetting aviation greenhouse gas emissions in a deep-cut carbon world? *Journal of Air Transport Management, 63,* 71–83.

Brundtland, G. H., Khalid, M., & Agnelli, S. (1987). Our common future. New York: Oxford University Press.

The Compact Edition of the Oxford English Dictionary. (1971). Oxford: Oxford University Press.

Greenair Online. (2019, March 15). *Agreement on sustainability criteria and advisory body for CORSIA carbon credits welcomed, although with reservations from NGOs.* Retrieved from https://www.greenaironline.com/news.php?viewStory=2577.

IATA. (n.d.). *Climate change.* Retrieved from https://www.iata.org/policy/environment/Pages/climate-change.aspx.

IATA. (n.d.). *IATA carbon offset program.* Retrieved from https://www.iata.org/whatwedo/environment/Pages/carbon-offset.aspx.

ICAO—International Civil Aviation Organization CORSIA Emissions Units Eligibility Criteria. (2019, March). Retrieved from https://www.icao.int/environmental-protection/CORSIA/Documents/ICAO%20document%2009.pdf.

International Council for Sustainable Aviation. (2019). *Letter from the International Coalition for Sustainable Aviation to the ICAO Council on the adoption of the Emissions Unit Criteria EUCs in order to address double counting and increase CORSIA's transparency.* Retrieved from https://www.edf.org/sites/default/files/documents/Letter_from_ICSA_to_the_ICAO_Council_members_regarding_the_Emissions_Unit_Criteria_27_Feb_2019_final.pdf.

Intergovernmental Panel on Climate Change. (2018). *Global warming of 1.5°C.* Retrieved from https://www.ipcc.ch/sr15/.

Ki-moon, B. (2010). *Pulling together in testing times: Securing a better future for all. Address to the 65th general assembly.* New York, NY: United Nations.

Laville, C. (2019, February 11). Critics attack secrecy at UN body seeking to cut global airline emissions. *The Guardian.* Retrieved from https://www.theguardian.com/.

Lyle, C. (2018). Beyond the ICAO's CORSIA: Towards a more climatically effective strategy for mitigation of civil-aviation emissions. *Climate Law, 8*(1–2), 104–127.

Martin, C., et al. (2016). *Oeko Institut, study prepared for DG CLIMA, how additional is the Clean Development Mechanism?* Reference: CLIMA.B.3/SERl2013/0026r. Retrieved from https://www.verifavia.com/uploads/files/clean_dev_mechanism_en.pdf.

Nam, U. V. (2015). Transforming our world: The 2030 agenda for sustainable. New York, NY: Division for Sustainable Development Goals.

Palmer, W. J. (2016). *Will sustainability fly?: Aviation fuel options in a low-carbon world* (pp. 43–46). London: Routledge.

Roundtable on Sustainable Biomaterials. (n.d.). *The RSB principles.* Retrieved from https://rsb.org/about/what-we-do/the-rsb-principles/.

Roundtable on Sustainable Biomaterials. (n.d.). *The RSB standard.* Retrieved from https://rsb.org/the-rsb-standard/.

Union of Concerned Scientists. (n.d.). *Global warming impacts, the effects of climate change are already here.* Retrieved from https://www.ucsusa.org/our-work/global-warming/science-and-impacts/global-warming-impacts.

CHAPTER 3

Boulding, Brundtland, Economics, and Efforts to Integrate Air Transportation Policies into Sustainable Development

Kenneth Button

> Anyone who believes that exponential growth can go on forever in a finite world is either a madman or an economist.
>
> Attributed to Kenneth Boulding (US Congress 1973)

> Development that meets the needs of the present without compromising the ability of future generations to meet their own needs.
>
> Brundtland Report (1987)

> The long run is a misleading guide to current affairs. In the long run, we are all dead.
>
> Maynard Keynes (1923)

K. Button (✉)
Schar School of Policy and Government,
George Mason University, Arlington, VA, USA
e-mail: kbutton@gmu.edu

T. Walker et al. (eds.), *Sustainable Aviation*,
https://doi.org/10.1007/978-3-030-28661-3_3

1 Introduction

The idea of sustainable development in the sense of continued economic and social improvement is certainly not new. Malthus (1798), for example, provided the classical economic argument about its virtual impossibility. But the concept, in a somewhat wider and more holistic sense, was resurrected some half-century ago not in the narrow Malthusian sense of physical constraints regarding food supplies, but in terms of the entire global resource base. This chapter is concerned with both the theoretical and the policy challenges arising from this more contemporary approach to sustainability, taking just one sub-sector of the economy, air transportation, as an example.

The chapter looks especially at the longer-term implication for aviation policies stemming from the work of the English economist Kenneth Boulding, and partly as an outcome of his work, the publication of the Brundtland Report (*Our Common Future*—hereafter just called "Brundtland") in 1987.[1] As seen in the first two opening quotations, both, albeit in different ways, Boulding and Brundtland directed a global audience to the challenges society faces over coming decades.[2] Many of the intellectual arguments put forward were not new, intellectually Malthus basically "has it all." Yet, Boulding's and Brundtland's arguments clearly articulate the idea of sustainability within a holistic framework of analysis and policy constraints. In turn, this has led to a more refined thinking both about the science of the environment and, from the perspective of this chapter, the way policies, and especially neoclassical-based economic policies, need adjusting. The arguments are set within the context of the rather less than the holistic market for air transportation, and in particular, the meaningful question of whether the notion of "sustainable transportation," or indeed sustainable air transportation, is raised.

Both Boulding and those involved in producing the Brundtland Report take a long-term, strategic view of economies. As the second of the opening quotations emphasizes, this fundamentally involves developments that meet the needs of the present without compromising the ability of

[1] For an overview of Boulding's work and approach, see, Keyfitz (2017).

[2] There were other influential reports and studies in the 1970s that were in a similar view to Boulding analysis, such as the Club of Rome's (1972) Limits to Economic Growth, but to make things manageable we focus our attention on just two strands of work.

future generations to meet their own needs.[3] This embraces not simply material resources, but also matters such as political acceptability and wider social well-being. Thus, it extends beyond the narrow, physical environmental considerations that previously dominated many of the economic debates on sustainability with their focus on matters such as GDP growth. Following Boulding and Brundtland, this chapter moves away from the rather naïve establishment and traditional neoclassical academic approaches that underlie such concepts as "sustainable transportation," "sustainable cities," and "sustainable housing" to focus on where transportation fits within a larger sustainable development scenario. In this sense, we are not concerned with any notion of sustainable aviation as such, but rather with whether current policies allow air transportation to fill an appropriate role within a larger, holistic sustainability framework.

So where do Maynard Keynes' well-known words set out in the third opening quotation fit into all this? On the surface, they seem to contradict the Boulding/Brundtland line of argument, but the verbiage's meaning is often mistaken. In fact, Keynes was saying that we should not recklessly enjoy the present and let the future go hang. He was concerned that the short-term economics needs of people should be accommodated within longer-term optimal growth paths. But he felt the main economic problems, and especially unemployment, were still unresolved in the 1930s and would take at least a century to be realized. While he wrote little specifically regarding the environment, Keynes' ideas are relevant to the concept of sustainable development.[4] They provide a basis for the other legs of the sustainability stool and especially highlight that the societal state needs to ensure living conditions to meet human welfare needs. In his opinion,

> The pace at which we can reach our destination of economic bliss will be governed by four things - our power to control population, our determination to avoid wars and civil dissensions, our willingness to entrust to

[3]Boulding (1966), for example, argues that "Economists in particular, for the most part, have failed to come to grips with the ultimate consequences of transition from the open to the closed earth." (p. 2) and that "The closed earth of the future requires economic principles which are somewhat different from those of the past" (p. 7).

[4]Readers are direct to Berr (2009) and Holt (2005) for arguments concerning Keynes's thinking on sustainable development. They also point to the relative neglect of this in most subsequent Post-Keynesian Economics.

> science the direction of those matters which are properly the concern of science, and the rate of accumulation as fixed by the margin between our production and our consumption. (Keynes 1930, p. 98)

Indeed, a full reading of Brundtland finds the majority of space devoted to the challenges Keynes' set down, and notably the handling of short-term problems of famine and economic deprivation.

Focusing more on the micro-level, sustainable development in the Boulding/Brundtland sense allows trade-offs between various activities and sectors with an emphasis on long-term resource retainment, i.e., about husbandry in the broadest sense. The chapter looks at the larger trends in transportation policy, and especially as it has affected aviation over the past 30 years. It considers the extent to which they meet the Boulding/Brundtland approach of maintaining an isolated planet in space and the wider political and economic Brundtland idea of sustainable development as opposed to much narrower, very incomplete, and possibly harmful concepts such as those of sustainable aviation, or even sustainable transportation. It is accepted that there has been progress in the ways that holistic development is viewed, and even if the ideas of both Boulding and Brundtland may now seem a little dated, the essence of their arguments remains at the core of policies aimed at the maintenance of the natural environment and of human societies that are still discussed and argued over.

The paper first considers the meaning of sustainable development in an economic sense and with a focus on the specifics of air transportation. It then moves on to examine the extent to which aviation policy has nudged the sector toward a more sustainable trajectory. This is done remembering that sustainable development is a holistic concept consistent with a greater use of resources in one sector provided there are at least comparable reductions elsewhere. In particular, it implicitly assesses the usefulness of the available battery of economic policy instruments in fulfilling this idea of trade-offs within the confines of finite overall resources.

2 Ideas of Sustainable Development

Definitions are important, and they can often vary over time. Going back to the medieval period in Western Europe and Asia, the idea that society should be interested in something called "sustainability" in the modern sense would have been thought rather strange. Minds were concentrated elsewhere. Between about 1270 and 1470, the Black Death was one of the most devastating pandemics in human history, resulting in an estimated 75–200 million deaths in Eurasia, and about 30–60% of Europe's population was affected. More widely, it reduced the world population from an estimated 450 million down to 350–375 million by the fifteenth century, and it took 200 years or so for economic recovery. It then recurred as outbreaks in Europe until the nineteenth century. Sustainability was thus largely thought off as not succumbing to global disease or staving because of natural disaster, rather than being a matter of sustaining a natural environment or resource base.

The Industrial Revolution saw this change. Although Malthus cast something of a shadow over things with his arguments of an underlying "Malthusian specter," and the inability of food production to expand at the pace of population growth, innovative ideas on the division of labor and discoveries of new natural resources seemed to offer road maps to sustained economic development. The economic ideas of Adam Smith and David Ricardo provided, respectively, the intellectual arguments for the division of labor and for free trade to carry this through.

The subsequent Victoria era, despite the age's outward confidence in its ability to sustain economic progress, had, however, an undercurrent of concerns about the long-term resource capacity to carry it through. The economist Stanley Jevons highlights this:

> Are we wise in allowing the commerce of the country to rise beyond the point at which we can long maintain it? ...After all, commerce is but a means to an end – the diffusion of civilization and wealth. To allow commerce to proceed until the source of civilization is weakened and overturned, is like killing the goose to get the golden egg. Is the immediate creation of wealth to be our only objective? Have we not hereditary possessions in our just laws, our free and nobly developed constitution, out rich literature and philosophy, incomparably above wealth, and which we are beyond all things bound to maintain, improve, and hand down in safety? And do we accomplish this duty in encouraging a growth of industry which must prove unstable, and perhaps involve all things in its fall? (Jevons 1865, p. 371)[5]

[5]There is also evidence of a considerable interest in sustainability by land-owners and colonists Europe at this time, although they were driven by commercial considerations rather than social ideas—see, for example, Jeremy Caradonna (2014).

The twentieth century concept of sustainable development as found in the likes of *Our Common Future* differs somewhat from that of the Victorian era, although there are practical similarities. The constant availability of carbon-based energy sources is one such common theme and one very pertinent for air transportation. Another element of modern sustainable thinking that perpetuates the ideas of nineteenth left-leaning thinkers, such as Friedrich Engels (1845), is its emphasis on the needs for appropriate political and social frameworks to ensure efficient use of resources and social justice. By the late twentieth century, enhanced knowledge, new technologies, and shifting geopolitical conditions had nuanced and expanded the concerns of the Victorian age.

Kenneth Boulding, with his analogue of earth being like a "space -ship," was one of the major intellectual economic driving forces in the second-half of the twentieth century. He was behind the shift in thinking toward the need to treat resource management within a closed system.

> I am tempted to call the open economy the "cowboy economy," the cowboy being symbolic of the illimitable plains and also associated with reckless, exploitative, romantic, and violent behavior, which is characteristic of open societies. The closed economy of the future might similarly be called the "spaceman" economy, in which the earth has become a single spaceship, without unlimited reservoirs of anything, either for extraction or for pollution, and in which, therefore, man must find his place in a cyclical ecological system which is capable of continuous reproduction of material form even though it cannot escape having inputs of energy. (Boulding 1966, p. 9)

Brundtland offers a pragmatic way of operationalizing the key sustainability concepts within a geopolitical framework.[6] The Report also goes somewhat further than Boulding in arguing that the environment goes beyond the physicality that forms the basis of most economic schools of thought to include social and political atmospheres and circumstances.

[6] In fact, even the Brundtland Report was not all embracing, but one of several commissioned by the UN, which also included Willy Brandt's (1980) study of the problems of inequality in the world and the failure of its economic system.

3 "Brundtland" and Its Immediate Aftermath

Brundtland is in many ways an extension of the concept of "The Great Transition," a term used by the Great Transition Initiative and its predecessor, the Global Scenario Group, to describe a vision of a just and sustainable global future. The term originates in Boulding's (1964) *The Meaning of the 20th Century—The Great Transition*. It describes the shift from pre-modern to post-modern culture, and the four possible courses of action that these organizations believe will allow humanity to successfully manage the Great Transition. Elements of this vision include egalitarian, social, and ecological values; increased inter-human connectivity; improved quality of life; and a healthy planet, as well as the absence of poverty, war, and environmental destruction, which are similar to Keyne's views. The practical challenge has always been how this was to be implemented.

To some extent, this led to Sweden suggesting in 1968 that the United Nations Economic and Social Council should have a conference focusing on human interactions with the environment. This was held in 1972 with a focus on "stimulating and providing guidelines for action by national government and international organizations" facing environmental issues (United Nations 1972). While there was considerable agreement on the need to tackle issues of poverty and the environment, little by way of concrete outcomes emerged. This theme was subsequently readdressed in a more pragmatic way in the Brundtland Report (Brundtland 1987), a UN emanated exercise from a World Commission on Environment and Development study headed by Gro Harlem Brundtland, later Prime Minister of Norway.

The Commission's targets in seeking a sustainable development path were multilateralism and interdependence of nations. *Our Common Future*, with its discussion of the environment and development in the widest sense as one single issue, placed environmental issues firmly on the political and economic agenda. A somewhat vague objective was set, namely to ensure "...development that meets the needs of the present without compromising the ability of future generations to meet their own needs" (Brundtland 1987). The aim being to maintain economic advancement and progress while protecting the long-term value of the environment. It was thus seen as providing, "... a framework for the integration of environment policies and development strategies" rather than a detailed plan for action. To achieve this, it was argued, requires the integration and acknowledgment of environmental, economic, and social concerns throughout the decision-making process. Much of the general and specific debates about achieving this, however, then moved onto institutional changes and physical regulation.

4 Where Does Air Transportation Fit into Sustainable Development?

To relate this general discussion of sustainability to aviation, we now move back a step and look at the specifics of air transportation in the context of its consumption of resources. Transportation in general is unquestionably a major contributor to climate-change related gas emissions, something over 25% of the global total, and it has proven particularly challenging to develop technologies that reduce these emissions without significantly reducing mobility (Pearce 2005). Air transportation contributes about three percent of climate-change gases, but its importance is growing and, while major technological advances are being made, the demands for its services are outpacing these. In addition, transportation is a network industry and there are very few movements that involve just one mode; indeed, walking may be the only one. For example, the air transportation supply chains for both personal and freight movements involve considerable terrestrial movements both to and from airports, as well as movements about the terminals.

The problems are also not just technical. Efficiency in mobility means a general lack of any natural barriers or at least the technical means of overcoming or circumventing them. Historically, humans have been very effective in doing the latter, and aviation, by offering super-terrestrial movement has particular physical advantages, and especially so in developing markets in Asia and Africa. Air transportation, however, also frequently involves movement across legal borders and frontiers. These are man-made constraints, often making institutional coordination of policies difficult, and their governance major challenges. Humans have not always been good at handling these challenges. But such coordination, either through markets or regulatory-reforms, is what is required in handling global challenges such as the emissions of climate-change gases or pandemics at the global level and local pollution such as NOx emissions or water pollution.

In addition to CO_2 emissions, it is well documented that the air transportation supply chain is responsible for a cocktail of other pollutants

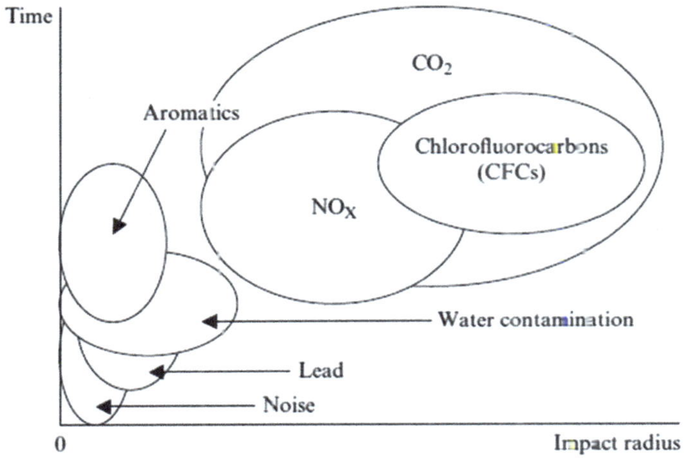

Fig. 1 The temporal and spatial coverage of exhaust gases and other environmental intrusions (*Source* Button 2019)

and contaminants with varied implications across time and space. Figure 1 provides a schema setting out the duration of the adverse effects of the main pollutants and contaminates together with the extent of their geographical spread from source.

There is no intention of detailing all the effects, but some are particularly important regarding aviation. Besides CO_2, aircraft contribute to climate change through the emissions of nitrogen oxides, which are particularly effective in forming greenhouse gas ozone when emitted at cruise altitudes. Accessing airports, especially using road transportation, has its implications. Lead, an "anti-knock" additive to gasoline to make engines run more smoothly, and a contributor to brain damage in children (McCubbin and Delucchi 1999) has relatively short-lived local impact terms of lingering in the environment. Lead's later replacement with aromatics involved introducing carcinogenic components. While often ignored in the aviation academic literature, such local pollution is an extremely important consideration in airport access. Even in Europe where there are generally modal options, automobiles dominate and carry 75% of traffic to and from airports. It rises to just under 90% at major US airports.

At airports, without adequate trapping, there can often be runoffs of aircraft deicing agents that contain significant amounts of glycol, as well as warm water, into the local watercourses that can create dead zones for aquatic life. It has long been recognized that aircraft noise is a major short-term issue at and around airports, not only because of its nuisance effects, but also by disrupting sleep, which can affect health and economic productivity (Flowerdew 1972).

One of the major problems given this diversity of effects, together with variety in their geographical spread, is to define some structure of subsidiarity to allot administrative responsibilities for handling each. A problem that is compounded when such as winds can lead to spread effects from the origin, and air transportation, and access to it, is by definition continually on the move. In the aviation context, there is considerable international travel which involves carriers from numerous countries taking off and landing at airports with differing rules, regulations, and charging policies, while being simultaneously guided by a plethora of national air traffic control systems, many of which are only just being able to communicate with each other. The existing institutional framework is neither an efficient market nor an efficient regulatory structure that allows an easy move to a system compatible with sustainable development.

Aviation has become more vital over time for both social cohesion and political stability. Historically, many of the major air transportation initiatives have been explicitly designed to increase political integration. The first major scheduled air services in the USA, for example, were primarily to carry long-distance mail, with the occasional passengers taken when there was capacity. Early major airlines, such as the British Imperial Airways, were created to tie an empire together (Lyth 2000), a theme that is continued at the super-macro level, for example, with the European Union's Trans-European Transport Networks initiative (Button 2011).

More recently, air transportation has become important in ensuring efficient use of resources and their distribution across diverse users; aviation now transports between 35 and 40% of world trade by value. Institutional interactions facilitated by air transportation have been recognized as socially, economically, and environmentally important at the global level. The World Bank, and the Asian, African, and American Development Banks support infrastructure investments, and international agencies such as the UN's International Civil Aviation Organization provide international "orgware"—the capacity to build the institutions that facilitates the efficient use of this infrastructure.

Combined, these factors lead, at the very least, to two overlapping general issues. First, the trade-offs between transportation and other activities in attaining sustainable development objectives, and second, within transportation, the trade-offs between the different impacts that various modes and their management have on the various components of a sustainable development scenario. Both issues involve consideration of institutional structures in the broadest sense.[7]

5 A Positive Role for Markets?

Despite the political attractions to many of large-scale public planning and government interventions aimed at sustainable objectives that were mooted almost immediately after Brundtland appeared, a number of economists emphasized that they were neither completely convinced that physical regulations nor an extensive battery of formal institutions were necessarily the best ways forward. In particular, they questioned the postulate of trade-offs between environmental sustainability and economic development, and argued in many cases that environmental protection can accompany economic development and is not inconsistent with profit motives. Many of these arguments go back to the ideas advanced sixty years or so earlier by Arthur Pigou (1920), and, in a somewhat different way, by Ronald Coase (1960) nearly three decades prior to Brundtland in terms of property rights.

In the case of Pigou, the cause of many of the excessive environmental costs of transportation are put down to the existence of negative externalities, costs that consumers of the environment neither shoulder nor react to. Following Pigou, they are the result of individuals or firms not being incentivized to pay the full costs of their actions because property rights are imperfectly assigned. If these sorts of views are accepted, then excessive environmental damage can be seen as an inefficient resource allocation that can be solved through improved production processes that in turn lead to enhanced competitiveness for those adopting them.[8] The role of the government in these cases according to Pigou is to impose prices that adjust the behavior of those generating the negative environmental externalities, or subsidies to encourage the use of less environmental intrusive technologies.

[7] Although not specific to transportation, Oliver Williamson (2000) provides details of the "New Institutional Economics" in which transportation policy may productively be assessed.

[8] This, for example, is the position taken by Michael Porter and Claas van der Linde (1995).

A major challenge from an intellectual perspective, and one important for the sustainability debate, is that Pigou's analysis is one of partial equilibrium that explicitly assumes the particular problem under review is set in a context whereby all other uses of the environment are optimized. This is perhaps a valid assumption in terms of some of the local transportation generated pollutants that were isolated in Fig. 1 but is not really realistic in the case of regional or global environmental concerns.

Another problem is the oft-made economic assumption that deviations from a sustainable development path are largely due to Pigouvian market failures, be they negative externalities or the existence of public goods. This is often true, but not always so, and certainly regarding air transportation, government intervention failures can be the problem (Organisation for Economic Cooperation and Development 1992; Button 2005; Winston 2006). These latter failures may, for example, be due to some form of capture of the system by parties with political or commercial influence seeking to serve their private interests rather than the interests of society (Stigler 1971), or be the result of poor understanding of issues, or inadequate data. Such failures are not uncommon in transportation, and aviation offers a good example of where this has been widespread in the past.

Air transportation has been heavily regulated since its inception, mainly for safety and strategic reasons. Economic regulations involving fare setting, market entry, and route configurations have also been prevalent in many domestic and international markets. The rationale here is often based on infant industry arguments that airlines needed to grow to a minimum size to be commercially viable, or that the industry is inherently monopolistic in the long term. Many of these markets were "deregulated" following the demonstration effects of 1977 (cargo) and 1978 (passenger) policy changes in the USA. While details obviously differ by country and routes, the implications of deregulation for aviation have been significant. Fares have fallen, and dynamic price discrimination is now the norm allowing much wider social strata to travel with resultant greater social and geographical cohesion. This has also led to much higher load factors. These have risen from just over 50% in the early 1970s in the USA to the mid-80% now. This effectively allows 60% more passengers per plane and thus significantly decreases the environmental footprint of each traveler. With fare setting, freedom airlines now compete more on fares and have become considerably more fuel price conscious. This in turn has led to the fuel efficiency of aircraft increasing by over one percent per year during the past 40 years (Button et al. 2015).[9] The problem is that traffic volume has also grown significantly.

The Coasian (1960) view is theoretically more holistic. Broadly speaking, the allocation of property rights to all resources and allowing owners to either trade them physically or buy and sell them in markets remove issues of estimating Pigouvian taxes or allocating the resultant revenues. The problems lie in the practical application. Challenges include how environmental costs will be measured, how initial allocations of the resources will be done, and the policing of the subsequent markets. Dealing with these problems would mean that ultimately the scarce nature of all resources leads markets to deploy them optimally over time.

If these types of markets were established for all forms of resources, they would produce outcomes close to the ideas generated in a Walrasian general equilibrium approach and avoid any need to make ceteris paribus assumptions (Walrus 1874). In practice, however, full allocation of property rights are unlikely. The cap-and-trade measures, for example, as used in the European aviation market and discussed below, move someway in the Coasian direction but are largely employed in isolation implicitly assuming other sectors meet sustainable criteria. In this sense, they are akin to the Pigouvian approach, although generally a more efficient instrument for meeting specific environmental targets. A particular challenge is to decide what the target, the "cap" should be. This inevitably depends on scientific knowledge and the actions being taken in all other sectors of the economy.

6 Possible Economic Instruments for Sustainable Development

There is a battery of instruments with an underlying economic rationale that can be used to mitigate the physical and social impacts of transportation. But how do they tie in with notions of sustainability and, in particular, how effectively have they been deployed in aviation? I list a few of the tools available that have been employed as illustrations of how attempts to treat transportation market imperfections in isolation often

[9]The outcome, while most agree has seen a social welfare gain, it has not produced a social optimal outcome. What reform has done is to also increase the amount of air transportation. Appropriate Pigouvian pricing, however, would act to influence the choices of individuals regarding the number of personal air trips they would be willing to pay for.

have unpredicted and sometimes perverse implications for sustainable development. But first a few words on the objectives and possible temporal conflicts in actions.

6.1 *Trade-Offs in Objectives*

There are trade-offs to consider and reducing one environmental effect of transportation can often increase another or introduce an additional one. This problem has been well documented in the surface transportation context which is highly relevant in terms of airport access by road. Reducing NOx emissions by fitting catalytic converters reduces fuel efficiency and thus increases CO_2 emissions. Additionally, as noted earlier, taking lead from gasoline often leads to the introduction of aromatics with carcinogenics being emitted. In a different way, changes in airline fleet composition are also important. In this context, Givoni and Rietveld (2010) find that the choice of service frequency and aircraft size has important environmental consequences. In particular, they find that the increase of aircraft size and service adjustment to offer similar seating capacity, as is often advocated, will increase local pollution but decrease climate change impact and noise pollution.

Transportation also plays an important role in the attainment of other objectives of sustainability, such as social cohesion, as was recognized in Boulding's work and discussed above. Transportation is often important in stimulating economic development, and almost without exception, always important in the distribution of the goods produced across populations. In this sense, there are inevitable trade-offs in virtually all initiatives to reduce the environmental intrusion of all forms of transportation, including aviation; it pushes up the latter's relative costs and affects various groups differently with those that are more transportation dependent being adversely affected.

6.2 *Dealing with Keynes' "We Are All All-Dead" Problem*

The challenges of forecasting the implication of current resource use into the future create ongoing problems in applying sustainable policies. Unlike a planned space flight, a lá Boulding, we are not exactly sure what the earth's resources are, and how their use and manipulation in the short term will impact their value for future generations. Economists usually approach this intergenerational issue, albeit it very imperfectly, by making use of social discount rates to reflect the relative importance of actions for future generations. This is done within a cost–benefits analysis (CBA) framework.

The use of discounting in the CBA arithmetic means that long-term costs and benefits are given relatively low weightings both at market rates and at the social discount rates normally adopted—the latter generally being somewhat below the market rates to reflect imperfections in the former. The UK's Stern (2008) Review of the implications of climate gas emissions, for which transportation is increasingly responsible, suggests that for economic and ethical reasons, much lower discount rates are appropriate—as low as 0.1%. This gives much more weight in the calculation of the net present value of a project to costs and benefits occurring in the more distant future. The Review has not been without its critics, and much of the discussion of the economic aspects of its findings, as opposed to the physical science, have revolved around the appropriate discount rate to deploy in developing sustainable development policies (see, e.g., Nordhaus [2007] and Weitzman [2007]).[10]

I am not concerned here with whether the Stern Review's approach to discounting is appropriate, but rather I accept that it is. The point then becomes, what general challenges does this pose for a CBA of any transportation project?

The measurement issues surrounding temporal valuations of costs and benefits effectively stay the same for most transportation projects, but different perhaps in form and the items to embrace. However, the almost equal temporal weights attached to the physical effects of a project, and ipso facto their equivalent monetary implications over a much longer period as argued by Stern, requires far more accurate long-term forecasts. This problem of handling longer-term forecasting is a generic challenge that must occur across many areas of public policy if the implications of the Stern Review are accepted, but I look at it specifically in the transportation policy context. The evidence here is that forecasts of costs and benefits are poor, often because effects on traffic flows are hard to predict. Flyvbjerg (2007) and Flyvbjerg et al. (2004), for example, give ample general examples of over-optimistic forecasting of usage and under forecasting of costs of supposedly more environmentally benign transportation modes.

[10] Heal (2017) offers an up-to-date review of the academic work in the field.

While there are certainly technical problems that cause considerable uncertainty, these sorts of problems arise in other sectors of the economy. Transportation, however, seems to be one sector where the biases of decision-makers are very influential, and this holds for all modes. One economic argument put forward by Flyvbjerg stems from the ideas of cognitive biases inherent in prospect theory (Kahneman and Tversky 1979). This theory argues that there are asymmetric attitudes in risk-taking, resulting in excessive optimism regarding some forms of decision-making to emerge. Much also seems to be due to prejudices on the part of decision-makers when selecting alternative forecasts and the incentives that motivate them (Button 2010).

6.3 Marketable and Tradable Permits

While theoretically the Coasian approach has intellectual merit, it has limitations. In terms of the prior example, the question arises as to whom property rights should be allocated in the first place. Ultimately, in a perfect market situation, trading of environmental resources should optimize their use. However, the Keynesian short-term political problem remains: Who gets the rents from the initial allocation? Markets are also unlikely to be perfect. For one thing, the quantification of environmental resources can be difficult. While this can be proxied in the case of something like greenhouse gas emissions by using tons of carbon, finding appropriate surrogates is difficult with implications like aircraft take-off noise or visual intrusion. There is also the issue of policing to ensure that contracts are upheld, and there are the transactions costs involved in the buying and selling of the rights themselves. There is also an assumption of no monopoly power being present in either side of the market.

Despite these challenges, tradable permits, or at least policies containing them as a main element, have been used to deal with specific transportation-related externalities, in addition to more generic problems such as global warming gas emissions. Tradable permits were, for example, part of the US policy to remove lead from gasoline, which had considerable environmental implications for airport road access pollution (Hahn 1989).

On an international scale, tradeable or marketable permits within cap-and-trade regimes have been gradually adopted since the Kyoto Protocol's acceptance in 1992. Carbon trading was at the forefront of this initiative, although the immediate aviation market was excluded. It was, however, belatedly in 2012 included in the European Union's

Emission Trading Scheme (EU-ETS) that initially only covered large factories, power stations, and other installations (European Commission 2006; Kopsch 2012). The academic analysis of the policy's implications produced a plethora of studies that largely concluded that it would, given the caps imposed and the price elasticities assumed, not have a disastrous overall effect on airline sector (e.g., Albers et al. 2009; Anger and Köhler 2010; Schaefer et al. 2010; Scheelhaase and Grimme 2007), although Barbot et al.'s (2014) analysis did suggest that the initial allocation mechanism would deter new airlines entering the market.

The inclusion of airlines within the EU-ETS, however, has proved problematic. Subsequent political and legal issues led to the European Commission to amend the Directive. It temporarily suspended the application of the ETS to airlines operating flights from or to non-EU destinations—a "stop the clock" derogation—in 2014. The airline industry and countries including China, India, Russia, and the USA had reacted adversely to the inclusion of aviation. They argued that the EU did not have jurisdiction to regulate flights when they were not in its members' airspace. In November 2012, the USA enacted the European Union Emissions Trading Scheme Prohibition Act that prohibits its carriers from participating in the ETS, despite the fact that enactment would seem to have minimal effect on its fleet (Malina et al. 2012), and China threatened to withhold $60 billion in outstanding airframe orders from Airbus.

At the global level, in October 2016 the International Civil Aviation Organisation (2016) implemented a quasi-market-based measure starting from 2021 to regulate international aviation emissions. The Carbon Offsetting and Reduction Scheme for International Aviation (CORSIA) aims to stabilize CO_2 emissions at 2020 levels, by requiring airlines to monitor their emissions on all international routes and buying offset credits to compensate the part of future emission growth above 2020 levels. The first two phases of CORSIA will run from 2021 until 2027 on a voluntary basis. In the interim, in December 2017, the European Commission extended until December 2023 the derogation from EU-ETS obligations for flights to and from third countries. Therefore, the EU-ETS only covers flights arriving in or departing from European Economic Area, regardless of the airline's nationality.[11]

[11] Some, such as Peeters et al. (2016), argue that the various, supposed technology difficulties that have prevented the wider introduction of carbon and other trading in environmental goods has, in fact, simply been a way for interest groups to protect their current position.

In terms of numbers, the original EU-ETS cap on aviation CO_2 allowances was 95% of 2004–2006 emissions levels, as specified in the EU Directive including aviation in the EU-ETS (221 million tonnes), although airlines could buy CO_2 allowances in the wider emissions market if they wished. The allowances are distributed as 82% is granted for free to aircraft operators, 15% to be auctioned, and three percent in a special reserve for distribution to fast-growing aircraft operators and new entrants. The allocation of allowances was scaled down from 2013 to 2023 to take account of the temporary reduction of the scope of the EU-ETS and flights between airports in the European Economic Area, with some upward adjustment to allow for Croatia joining the Union. Environmentally, while verified CO_2emissions from aviation activities carried out between airports in the EEA have increased from 53.5 million tonnes of CO_2 in 2013 to 64.2 million tonnes in 2017, the inclusion of intra-European flights in the EU-ETS has effectively delivered around 100 million tonnes of CO_2 reductions/offsets between 2012 and 2018 (Nava et al. 2018).

6.4 *Pigouvian Taxes and Subsidies*

Pigou's idea that either a tax on a negative environmental externality or a subsidy to a positive on poses problems. There are often confusions over what they are applied to. For example, public transit subsidies aimed at transferring traffic from automobiles to less environmentally intrusive modes are not Pigouvian subsidies; they do not directly affect the pollution per se. A direct payment to automobile drivers to reduce their driving is a Pigouvian charge. But on the other hand, a subsidy to encourage the use of high-speed railways rather than flying is not a Pigouvian subsidy, whereas a direct payment to air travelers to discourage them to fly or to airlines to use more efficient aircraft, are Pigouvian subsidies.[12]

Pigouvian policies also involve distributional matters. While a tax impacts directly on the polluter, a subsidy to discourage pollution is financed from general taxation or a hypothecated tax on those polluted. The Organisation for Economic Cooperation and Development's (1975) "polluter -pays-principle," for example, put the onus of emissions reduction on the polluter, but this is a normative approach rather than based upon any strict efficiency criterion.

[12] Lingeflyg merged with SAS in 1993.

The idea of environmental charging is now about a century old, and there have been attempts to introduce such policy into aviation (Alamdari and Brewer 1994). Fuel taxation, as a proxy for taxing carbon and other emissions, is perhaps the most widely discussed of them. For example, the Swedish government imposed taxes in 1989 on domestic flights at a rate of 12 Swedish krona per km of nitrogen oxides and hydrocarbons and 0.25 per km of carbon dioxide. In 1993, the CO_2 tax rate was increased to 0.32 krona. The emission taxes were based on the average Linjeflyg (the main carrier at the time) emissions of a flight of 380 kms (Elofsson et al. 2018).

At the more local level, the issue of noise nuisance has been long-standing, although reduced somewhat in recent years by improved engineering of aircraft. Fiscal tools have been used in many cases to somewhat mitigate this problem, although physical command-and-control measures, such as night curfews, bans on certain types, or gauge of aircraft, and flight path specifications, are more common. London's Heathrow Airport offers an example of landing fees differentiated by noise, with significantly higher fees at night and differentiation by aircraft type.[13] Similar types of differentiation are operative in most other major EU airports.

Additional to the use of Pigouvian style taxes, monetary values of noise nuisance have been included in most assessments of new airport construction and the expansion of existing capacity. This has a common practice since the work assessing the merits of sites for a third London airport in the late 1960s (*Commission on the Third London Airport* 1971). This approach obviously affects the scale and types of aviation activities at individual airports.

6.5 What Are the Implications of All This?

At least two things emerge from these accounts. The first is that there is a battery of policy tools available to move the environmental footprint

[13]The ICA0 divides aircraft into—Chapter 1: aircraft with type license before 1970 (e.g. B707s); Chapter 2: aircraft with type license between 1970 and 1978 (e.g. B747-100/200s, certain B737-200s) and Chapter 3: aircraft with type license after 1978 (e.g. A310, B757, B767, MD83). The UK system, following EU rules, limits flights to Chapter 3 aircraft and imposes a common fee with deviations from this according to specific aircraft type.

of air transportation toward the optimal position. But equally, none of these tools are perfect and most meet with political and public resistance because individuals neither like being constrained in what they do, nor to "pay more" for something previously seen as free. The second implication that emerges is that, given the intellectual and practical challenges of adopting these instruments in isolation, it becomes even more challenging when there are interacting, and often conflicting, social and political objectives to consider. Perhaps a third factor overrides the two previous ones, namely that the policy instruments, even cap-and-trade regimes, are partial equilibrium and hardly interface with other sectors of the economy. Their approaches involve variant of traditional neoclassical economics, with its ceteris paribus assumption, rather being holistic in orientation.

I now move to consider the difficulties of meeting these challenges in the context of current institutional structures.

7 The Challenges of Government and Governance

The overriding problem in realizing sustainable development extends beyond that of making transportation or any other activity environmentally benign. It involves not only issues regarding the usefulness of the available policy tools for meeting sustainable environmental objectives, but also the institutional structure available for carrying through overall sustainable development. Governments set laws and regulations, while bureaucrats are responsible for the governance involved in interpreting, prioritizing, and carrying through the laws and regulations (Williamson 2000). The problem is that for practical reasons this has entailed a stove-piping of decision-making with governments being comprised of ministries or cabinets with their own bureaucracies for overseeing interpretations, implementation, and enforcement of laws and regulations for specific sectors; e.g., Departments of Transportaion, Agriculture, and Energy.

Although the subject deviates from the core economic debates on sustainability—it is generally seen more as a matter of public administration—the structure of government and governance provides the institutional background against economic policies are set and implemented. Historically, a ministry of transportation has emerged in most countries with its own hierarchy of specialized bureaucrats.

In the aviation case, within ministries there are national Civil Aviation Administrations (CAAs) with designated duties regarding the industry.[14] It is assumed that entities such as this have the objective of serving the "public interest."

The latter view is often somewhat misguided given the work on institutional and behavioral economics. The existence of markets for power and influence, and the commonality of coalitions are common deviations in this context. Each ministry competes with other ministries for resources and bureaucrats, while occasionally serving the public interest. There is evidence that they often favor economic policies that serve their own agenda or those constituents to whom they are beheld to.[15] One way of looking at this from the public administration perspective is in terms of the fundamental tension about how governments should intertwine and balance professional policy knowledge with democratic decision-making. The form of the overarching institutional structure adopted, and the placement of an issue, such as the role of air transportation in a sustainable development strategy within it, can translate into significant differences in responsiveness and technical objectiveness.

In this context, there are questions of whether sustainability is treated as the responsibility of the executive or of a line department, such as transportation or the CAA, and, linked to this, whether there is a specialized unit within the government that is explicitly responsible for sustainability. The combination of the multifaceted and novel nature of the modern sustainable development concept, combined with the inertia inherent in the structures of government and governance, tends to stymie significant changes in the way institutions are constructed or interact. The outcome is that individual departments continue to focus on their specific interests with limitations in the ways that they coordinate to achieve sustainability objectives. Where there is an overarching "Department of Sustainable Development," it is unlikely to have the specialism to develop detailed policies, and is seldom given the status of, say, the Finance Ministry.

What research seems to show is that agencies that are strongly buffered from political influence produce more accurate and less biased

[14] These, at the very least, need to conform with ICAO requirements.

[15] As George Stigler (1971, p. 3) puts it, regulators do not solely advance the overall public interest by correcting market failure but instead regulation is "...acquired by the industry and is designed and operated primarily for its benefit."

policy information than agencies that have weaker institutional autonomy. In other words, they take a longer-term view of policies than elected politicians, and this is more in line with the needs for sustainability, always of course keeping in mind the words of Maynard Keynes concerning the long-run cited earlier.

8 Conclusions

The paper has, first, set out what was broadly advocated some 50 years ago by Boulding and later the Brundtland Report in terms of developing viable strategies to ensure the long-term future of the planet. Secondly, its has related these to the achievements in air transportation economics and policies that have followed this thinking. Movement forward has clearly been difficult.

First, it is clear that the debates surrounding the global environment in its broadest sense are far from over. These are not only about the highly vocal and visible fights over whether current global climate changes are largely due to human action but also embody wider social priorities over income distribution and political priorities. Even if the quantification of effects could be agreed upon, policy responses are unlikely to emerge with universal approval. The political response is likely to be replicating much of what has gone before to avoid making painful trade-offs. The equity-efficiency argument is an old one, but one that still dominates sustainable development policy, and within this, sectorial trade-offs.

Second, when reading the material in the general transportation literature and the more specific journals focusing on air transportation and the environment, one notices that there has been a massive increase in the formalizing of the issues involved in defining a sustainable development path. Coupled with this is the use of the ever-increasing amounts of data available for empirical analysis that helps put the various elements of the sustainability equation in perspective. This has in some ways led to a more holistic thinking about interactions involving the role of air transportation within wider sustainable transportation development scenarios, but this is still largely done in isolation from other sectors. Stove piping like this does not produce sustainable outcomes.

Third, the institutional structures are designed for an age when the stovepiping of policies was necessary and probably efficient given the information available and arguably simpler ideas of how the world works. Transportation departments, for example, made sense then. In the information age, and with much broader ideas of what constitutes development, the old administrative design does not cope well. Public policy in general needs reform to define a path to sustainable development, and sectors like aviation need to be integrated within this rather than treated in isolation.

References

Alamdari, F. E., & Brewer, D. (1994). Taxation policy for aircraft emissions. *Transport Policy, 1,* 149–159.

Albers, S., Bühne, J. A., & Peters, H. (2009). Will the EU-ETS instigate airline network reconfigurations? *Journal of Air Transport Management, 15,* 1–6.

Anger, A., &. Köhler, J. (2010). Including aviation emissions in the EU-ETS: Much ado about nothing? A review*Transport Policy, 17,* 33–46.

Barbot, C., Betanco, O., Socorro, P. M., & Viecens, F. M. (2014). Trade-offs between environmental regulation and market competition: Airlines, emission trading systems and entry deterrence. *Transport Policy, 33,* 65–72.

Berr, E. (2009). Keynes and sustainable development. *International Journal of Political Economy, 38,* 22–38.

Boulding, K. E. (1964). *The meaning of the 20th century: The great transition.* Lanham: University Press of America.

Boulding, K. E. (1966). The economics of the coming spaceship earth. In H. Jarrett (Ed.), *Environmental quality in a growing economy* (pp. 3–14). Baltimore: Resources for the Future.

Brandt, W. (1980). *North-South: A programme for survival—Report of the independent commission on international development issues.* Cambridge, MA: MIT Press.

Brundtland, G. H. (Ed.). (1987). *Our common future: The world commission on environment and development.* Oxford: Oxford University Press.

Button, K. J. (2005). Market and government failures in transportation. In K. J. Button & D. Hensher (Eds.), *Handbook of transport strategy, policy and institutions* (pp. 11–28). Oxford: Pergamon.

Button, K. J. (2010). Does the primary problem of applying cost-benefit analysis to transportation policies lie in the methodology or in the forecasts used? In E. Van de Voorde & T. Vanelslander (Eds.), *Papers in honor of Gust Blauwens.* De Boeck: Antwerp.

Button, K. J. (2011). Transport policy. In A. M. El-Agraa (Ed.), *The European Union: Economics and policies* (9th ed., pp. 244–256). Cambridge: Cambridge University Press.

Button, K. J. (2019). *Transport economics* (4th ed.). Cheltenham: Edward Elgar.

Button, K. J., Brugnoli, A., Martini, G., & Scotti, D. (2015). Economic factors affecting the registration of lower CO_2 emitting aircraft in Europe, *Transportation Research D, 38*, 117–124.

Caradonna, J. L. (2014). *Sustainability: A history*. Oxford: Oxford University Press.

Club of Rome. (1972). *Limits to economic growth*. Washington, DC: Potomac Associates.

Coase, R. H. (1960). The problem of social cost. *Journal of Law and Economics, 3*, 1–44.

Commission on the Third London Airport Report. (1971). London: HMSO.

Elofsson, A., Smedby, N., Larsson, J., & Nässén, J. (2018). Local governance of greenhouse gas emissions from air travel. *Journal of Environmental Policy & Planning, 20*, 578–594.

Engels, F. (1845). *Die Lage der arbeitenden Klasse in England* (Reproduced by HardPress Publishing, 2013).

European Commission. (2006). *Proposal for a Directive of the European Parliament and the Council amending Directive 2003/87/EC as to include aviation activities in the scheme for greenhouse gas emission allowance trading within the community*. Brussels: European Commission, COM(2006)818 final.

Flowerdew, A. D. J. (1972). The cost of airport noise. *Journal of the Royal Statistical Society, Series D, 21*, 31–46.

Flyvbjerg, B. (2007). Cost overruns and demand shortfalls in urban rail and other infrastructure. *Transportation Planning and Technology, 30*, 9–30.

Flyvbjerg, B., Holm, M., & Buhl, S. (2004). What causes cost overrun in transport infrastructure projects? *Transport Reviews, 24*, 3–18.

Givoni, M., & Rietveld, P. (2010). The environmental implications of airlines' choice of aircraft size. *Journal of Air Transport Management, 16*, 159–167.

Hahn, R. W. (1989). Economic prescriptions for environmental problems: How the patient followed the doctor's orders. *Journal of Economic Perspectives, 3*, 95–114.

Heal, G. (2017). The economics of the climate. *Journal of Economic Literature, 55*, 1046–1063.

Holt, R. (2005). Post-Keynesian economics and sustainable development. *International Journal of Environment, 1*, 174–186.

International Civil Aviation Organisation. (2016). *Resolution A39-2: Consolidated statement of continuing ICAO policies and practices related to environmental protection—Climate change*. Montreal: ICAO 39 Triennial Assembly.

Jevons, W. S. (1865). *The coal question; An inquiry concerning the progress of the nation, and the probable exhaustion of our coal mines* (1st ed.). London: Macmillan.

Kahneman, D., & Tversky, A. (1979). Prospect theory: An analysis of decision under risk. *Econometrica, 47*, 263–291.

Keyfitz, N. (2017). Kenneth Ewart Boulding: January 18, 1910–March 18, 1993. In *Biographical memoirs* (Vol. 69, pp. 3–14). Washington, DC: National Academy of Sciences.

Keynes, J. M. (1923). *A tract on monetary reform*. London: Macmillan.

Keynes, J. M. (1930). Economic possibilities for our grandchildren. *The Nation & Athenaeum*, Part I, 11 October, pp. 36–37 and Part II, 18 October, pp. 96–98.

Kopsch, F. (2012). Aviation and the EU emissions trading scheme—Lessons learned from previous emissions trading schemes. *Energy Policy, 49*, 770–773.

Lyth, P. (2000). The empire's airway: British civil aviation from 1919 to 1939. *Revue Belge de Philologie et D'Histoire Année, 78*, 865–887.

Malina, R., McConnachie, D., Winchester, N., Wollersheim, C., Paltsev, S., & Waitz, I. A. (2012). The impact of the European Union emissions trading scheme on US aviation. *Journal of Air Transport Management, 19*, 36–41.

Malthus, T. R. (1798). *An essay on the principle of population as it affects the future improvement of society, with remarks on the speculation of Mr. Goodwin, M. Condorcet and other writers*. London: J. Johnson in St Paul's Church-yard.

McCubbin, D. R., & Delucchi, M. A. (1999). The health costs of motor-vehicle-related air pollution. *Journal of Transport Economics and Policy, 33*, 253–286.

Nava, C. R., Meleo, L., Cassetta, E., & Morelli, G. (2018). The impact of the EU-ETS on the aviation sector: Competitive effects of abatement efforts by airlines. *Transportation Research Part A, 113*, 20–34.

Nordhaus, W. D. (2007). A review of the "Stern review on the economics of climate change". *Journal of Economic Literature, 45*, 686–702.

Organisation for Economic Cooperation and Development. (1975). *The polluter pays principle*. Paris: OECD.

Organisation for Economic Cooperation and Development. (1992). *Market and government failures in environmental policy: The case of transport*. Paris: OECD.

Pearce, D. (2005). The social cost of carbon. In D. Helm (Ed.), *Climate-change policy* (pp. 99–133). Oxford: Oxford University Press.

Peeters, P., Higham, J., Kutzner, D., Cohen, S., & Gössling, S. (2016). Are technology myths stalling aviation climate policy? *Transportation Research Part D, 44*, 30–42.

Pigou, A. (1920). *The economics of welfare*. London: Macmillan.

Porter, M. E., & van der Linde, C. (1995). Toward a new conception of the environment-competitiveness relationship. *Journal of Economic Perspectives, 9*, 97–118.

Schaefer, M., Scheelhaase, J., Grimme, W., & Maertens, S. (2010). The economic impact of the upcoming EU emissions trading system on airlines and EU member states—An empirical estimation. *European Transport Research Review, 2,* 189–200.

Scheelhaase, J. D., & Grimme, W. G. (2007). Emissions trading for international aviation—An estimation of the economic impact on selected European airlines. *Journal of Air Transport Management, 13,* 253–263.

Stern, N. (2008). The economics of climate change. *American Economic Review, 98,* 1–37.

Stigler, G. J. (1971). The theory of economic regulation. *Bell Journal of Economics and Management Science, 2,* 3–21.

United Nations. (1972). *Report of the United Nations conference on the human environment.* New York: United Nations.

US Congress. (1973). *Energy reorganization act of 1973: Hearings, ninety-third congress, first session, on H.R. 11510,* Washington, DC: US Government.

Walrus, M-.E-.L. (1874). *Éléments d'économie politique pure, ou Théorie de la richesse sociale.* Lausanne: Corbaz.

Weitzman, M. L. (2007). A review of the "Stern review on the economics of climate change". *Journal of Economic Literature, 45,* 703–724.

Williamson, O. E. (2000). The new institutional economics: Taking stock, looking ahead. *Journal of Economic Literature, 38,* 595–613.

Winston, C. (2006). *Government failure versus market failure.* Washington, DC: The Brookings Institution.

CHAPTER 4

Environmentally Sustainable Air Transport: A Focus on Airline Productivity

Thomas Budd, Mario Intini and Nicola Volta

1 Background and Context

Tackling air transport's environmental externalities represents one of the key challenges facing the industry. Airlines, airport operators, aircraft manufacturers, air navigation service providers, and regulators are increasingly coming to the realization that environmental impacts are key constraints on growth, and that sound environmental and commercial business practices are not necessarily mutually exclusive.

Like many other sectors, the framing of environmental protection measures in air transport is generally couched in terms of conforming to the guiding principles of sustainability, which is the long-held idea

T. Budd · N. Volta
Centre for Air Transport Management, Cranfield University,
Cranfield, UK
e-mail: t.m.budd@cranfield.ac.uk

N. Volta
e-mail: n.volta@cranfield.ac.uk

M. Intini (✉)
Department of Economics, Management and Business Law,
University of Bari Aldo Moro, Bari, Italy

T. Walker et al. (eds.), *Sustainable Aviation*,
https://doi.org/10.1007/978-3-030-28661-3_4

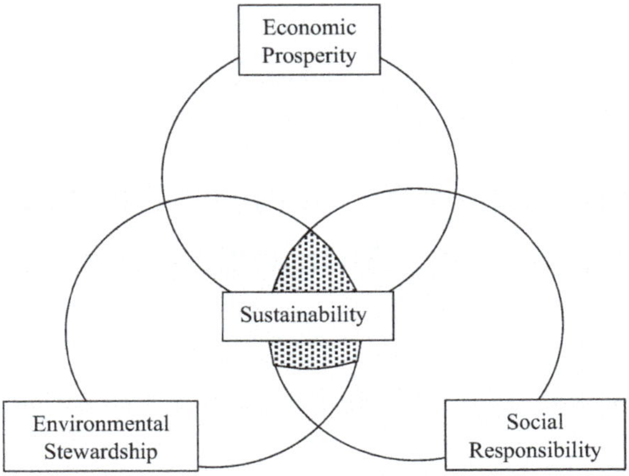

Fig. 1 The 'triple-bottom line' of sustainability (*Source* Adapted from Landrum and Brown, Inc. 2012)

of policies balancing environmental, social, and economic priorities in a way that meets the needs of present-day societies without jeopardizing the ability of future societies to do likewise. Sustainable practices are considered to be those that reduce or limit the environmental impacts of a process or practice, while at the same time creating financial benefits and social opportunities. Collectively, these aspects of sustainability are known as the 'triple bottom line' (see Fig. 1).

While the environmental impacts of aviation have long been understood (see Lee et al. 2009), political debates surrounding aviation's impact on the environment remain contested and contentious. Such debates are relevant in a global context due largely to the significant predicted near-to-medium-term growth in the global air transport market and the knock-on impacts this will have in terms of the environment. ICAO (International Civil Aviation Organization) forecasts a near tripling of the global commercial aircraft fleet, rising from just under 20,000 aircraft in 2010 to 56,500 in 2040, with accompanying increases in the passenger and air cargo markets (ICAO 2016). Concurrently, international aviation CO_2 emissions are forecasted to grow, under the 'baseline growth scenario including fleet renewal, from around 500 Mt in 2010 to nearly 2000 Mt in 2040' (ICAO 2016).

While recent years have seen a number of successful test flights using non-carbon-based biofuels, the current commercial air transport fleet relies almost exclusively on carbon-based fossil fuels (mostly kerosene). When burnt in aircraft engines, this process produces emissions of carbon dioxide (CO_2), nitrogen oxides (NO_x), particulates, water vapour, and other pollutants that alter the chemical composition of the atmosphere and contribute to climate change. Air transport is unique in that the majority of emissions are released at altitude where their impact is typically greater than if they were released at ground level.

Poor local air quality is also an issue in terms of the health of communities living around airports, with gaseous emissions of NO_x, carbon monoxide (CO), and particulates in particular a problem. As a result, tighter regulatory controls on standards of local air quality have been imposed in Europe and North America.

Aside from atmospheric emissions, local airport communities are also affected by noise from aircraft in flight or on the ground. The issue of aircraft noise has become central to issues of airport expansion, representing a key barrier to growth and the single environmental issue most likely to provoke action and complaint from airport communities (Budd 2017). Prolonged exposure to aircraft noise has been attributed to a host of adverse psychological and physiological effects, including increased blood pressure and sleep disturbance, as well as discomfort and depression. This situation persists despite continuous improvement in the design and manufacturing of engines and airframes over the past few years to make aircraft increasingly quieter. However, research has shown that it is the subjective perception of noise that is important (which is notoriously difficult to measure), rather than the objective acoustic level of the events themselves (Suau-Sanchez et al. 2011).

The environmental impacts of air transport also extend to the energy, water, and waste management issues. Airports, especially, are typically large consumers of electrical energy because of the maintenance of ambient air temperature (both heating and cooling), air quality, and lighting in the terminal building and across the airfield. Airports are also large consumers of water, used for drinking, catering, cleaning, and other purposes (Budd 2017). Air transport operations also produce large quantities of waste, ranging from solid waste from retail and in-flight catering operations to hardcore and specialist waste from construction and maintenance activities at the airport.

Although initiatives to address the environmental impacts from aviation have so far largely fallen into the category of voluntary efforts on the part of industry stakeholders, there are also increasingly more stringent regulatory requirements for stakeholders to address such issues. When it comes to emissions, the Kyoto Protocol, signed in 1997, was one of the first international agreements committing nations to set binding targets for emissions reductions. However, aviation (or indeed transport generally) was not included in the agreement, and the responsibility for environmental protection in aviation was left to ICAO. Between 1990 and 2012, emissions from international aviation grew by 75%. However, recent years have seen significant progress in an international regulatory context, with the ongoing CORSIA (Carbon Offsetting and Reduction Scheme for International Aviation) scheme representing the first widely ratified global market-based measure designed to tackle emissions from aviation (ICAO 2016).

As an industry, the airline business is notorious for its highly competitive nature and fine margins between success and failure. Productivity is a key success factor in aviation, and economic sustainability has traditionally been *the* central goal of the airline business. However, airlines are also increasingly having to account for environmental and social priorities in their business models without unduly jeopardizing their economic standing. Aside from the benefits accrued by the airlines themselves, a growing and prosperous air transport sector can also play an important catalytic role in the wider economy. As stated by IATA (International Air Transport Association), while air traffic demand has increased as economies have grown, air transportation itself can be a key cause and facilitator of economic growth. Greater connections to the global air transport network can increase productivity and growth of economies by providing better access to markets, improving links within and between businesses and providing greater access to international capital markets and resources. In 2017, IATA[1] noted that consumers benefited from lower 'real' travel costs and spent the equivalent of 1% of the world's GDP on air transport.

[1] https://www.iata.org/publications/economics/reports/industry-econ-performance/iata-economic-performance-of-the-industry-end-year-2017-report.pdf.

2 Aim of the Research

Our work aims to build on this important area of research by estimating the trend in airline productivity when considering both the economic and environmental aspects of air transport. We apply a data envelopment analysis based on the productivity index to estimate the economic evolution of the major global airlines. More precisely, we adopt a biennial Malmquist–Luenberger index (BML), which is an indicator of productivity that changes over time.

The peculiarities of our approach can be summarized as follows:

i. Both economic and environmental aspects are evaluated and described in a unique index.
ii. Non-parametric frontier technologies are used in computing the index.
iii. It is possible to decompose the index into technical efficiency and technology change.

Moreover, the indirect advantage of this approach is that in using non-parametric technologies (i.e., DEA), there is no need to impose an a priori functional form on the production function. From a more technical point of view, the adoption of a 'biennial' approach solves the problem of infeasibility under variable returns to scale and the non-identification of technical regress. Furthermore, it avoids the recalculation of indices when adding new time periods. By considering both positive (i.e., economical) and negative (i.e., pollution) externalities, we are able to describe the airline's path towards a more sustainable industry.

The following section details the methodology used in the chapter (Sect. 3). This is followed by an assessment of data trends (Sect. 4) and presentation of the results and a discussion (Sect. 5). Various conclusions from the analysis and some concluding remarks are provided at the end of the chapter (Sect. 6).

3 Methodology

In this section, we describe the methodology adopted in order to compute the economic and environmental productivity index. Section 3.1 describes the BML index as well as the frontier models applied. Section 3.2 describes the data collection and the airline CO_2 estimations.

3.1 The Biennial Malmquist–Luenberger Index

Our methodology approach follows the one proposed by Scotti and Volta (2015). The production possibility set (PPS) for $k=1, \ldots, K$ airlines producing M desirable outputs, $y=1, \ldots, M$ with $y \in R_{+}^{M}$, J undesirable outputs, $b=1, \ldots, J$ with $b \in R_{+}^{J}$, by using N inputs, $x=1, \ldots, N$ with $x \in R_{+}^{N}$, is represented by $P(x)$. Consequently, the PPS can be expressed as follows:

$$P(x) = \{(y, b)|x \text{ can produce } (y, b)\}, \tag{1}$$

As in Färe et al. (2007), we present the following axioms required to define our production technology:

$$\{0\} \in P(x) \text{ for all } x \in R_{+}^{N}, \tag{2a}$$

$$P(x) \text{ is compact for all } x \in R_{+}^{N}, \tag{2b}$$

$$P(x) \subseteq P(x') \text{ if } x' \geq x. \tag{2c}$$

These axioms define that (2a) inactivity is always possible, (2b) finite amount of inputs can only produce finite amount of outputs, and (2c) inputs are freely disposable. Regarding the undesirable outputs production, further axioms are required:

$$(y, b) \in P(x) \text{ and } 0 \leq \theta \leq 1 \text{ imply} (\theta y, \theta b) \in P(x), \tag{2d}$$

$$(y, b) \in P(x) \text{ and } y' \leq y \text{ imply } (y', b) \in P(x), \tag{2e}$$

$$\text{if } (y, b) \in P(x) \text{ and } b = 0 \text{ then } y = 0. \tag{2f}$$

Axiom (2d) defines a technology with weakly disposable undesirable outputs. Thus, reductions in undesirable outputs are possible if desirable outputs are reduced in proportion, or, in other words, it is costly to reduce undesirable outputs. θ can be interpreted as the abatement factor scaling down the activity level. Axiom (2e) defines a strong disposability of good outputs meaning that if an output vector is feasible, then any output vector with less desirable output is also feasible. Finally, according to axiom (2f), if desirable outputs are produced then some undesirable outputs will also be produced. Adler and Volta (2016) discuss current approaches and the respective assumption when modelling negative externalities in a production framework.

Once the production possibility set and the relative assumption are defined, it is possible to compute the distance of the units to the relative production frontier. A directional distance function (DDF) seeks to increase the desirable output while simultaneously decreasing the undesirable outputs. Formally, it can be defined as follows:

$$\vec{D}\big(x, y, b; g_y, g_b\big) = \max\big\{\beta \big| \big(y + \beta g_y, b - \beta g_b\big) \in P(x)\big\}, \tag{3}$$

where $g = \big(g_y, g_b\big)$ is a direction vector, with $g \in R_+^M \times R_+^J$, defining the direction of the increase (decrease) of desirable (undesirable) outputs. In the formulations, we will omit the direction vector $g = \big(g_y, g_b\big)$ in DDFs formulations (i.e., we will refer to $\vec{D}\big(x, y, b; g_y, g_b\big)$ as $\vec{D}(x, y, b)$). In order to define a BML index, we start with the definition of contemporaneous benchmark technology. Given an observation period $t=1, \ldots, T$, the contemporaneous technology is defined as follows:

$$P^t(x^t) = \{(y^t, b^t)|x^t \text{ can produce } (y^t, b^t)\}. \tag{4}$$

In Eq. (4), the production set is only made by the observation at time t. Considering the technology associated with the subsequent period $t+1$, we can define:

$$P^{t+1}(x^{t+1}) = \Big\{\Big(y^{t+1}, b^{t+1}\Big)\Big|x^{t+1}\text{can produce}\Big(y^{t+1}, b^{t+1}\Big)\Big\}. \tag{5}$$

Similarly, in Eq. (5), the production set is only made by the observation at time $t+1$. As in Pastor et al. (2011), the base t biennial P^B technology can be defined as the convex hull of period t and period $t+1$ technologies:

$$P^B = \text{conv}\Big\{P^t, P^{t+1}\Big\}. \tag{6}$$

Thus, a series of $T-1$ overlapping biennial technologies exist for each pair-wise comparison of adjacent periods. As stated in Pastor et al. (2011), two adjacent time periods are sufficient to establish the properties in order to avoid infeasibility, to allow for technical regress, and to maintain previous productivity computations. Starting from the above biennial technology and following Chung et al. (1997), we can define the BML as follows:

$$\text{BML}\left(x^t, y^t, b^t, x^{t+1}, y^{t+1}, b^{t+1}\right) = \frac{1 + \vec{D}^B\left(x^t, y^t, b^t\right)}{1 + \vec{D}^B\left(x^{t+1}, y^{t+1}, b^{t+1}\right)}, \tag{7}$$

where the DDF $\vec{D}^B(x, y, b)$ is defined on the benchmark technology defined in Eq. (6) and computed as in Eq. (3) for the adjacent time periods t and $t+1$. A BML index > (<) 1 means that the production enables more (less) desirable outputs and less (more) undesirable outputs for a given amount of inputs. Consequently, a BML=1 implies the absence of productivity growth. Moreover, since the technology includes both period t and $t+1$ technologies, we do not need to apply the usual geometric mean in defining Eq. (7). The BML index can be decomposed into the components of productivity growth as follows:

$$\text{BML} = \text{EC} \times \text{TC}, \tag{8}$$

where EC is the efficiency change and TC is the technical change. The former represents the efficiency gain over the time period, and the latter accounts for technical progress in the airlines' production function. The efficiency change is defined as follows:

$$\text{EC} = \frac{1 + \vec{D}^t\left(x^t, y^t, b^t\right)}{1 + \vec{D}^{t+1}\left(x^{t+1}, y^{t+1}, b^{t+1}\right)}. \tag{9}$$

As usual, EC > (<) 1 means an increase (decrease) in the efficiency of the decision making unit (DMU) with respect to the contemporaneous frontiers. The technical change factor is defined as follows:

$$\text{TC} = \left[\frac{\left(1 + \vec{D}^B\left(x^t, y^t, b^t\right)\right) / \left(1 + \vec{D}^t\left(x^t, y^t, b^t\right)\right)}{\left(1 + \vec{D}^B\left(x^{t+1}, y^{t+1}, b^{t+1}\right)\right) / \left(1 + \vec{D}^{t+1}\left(x^{t+1}, y^{t+1}, b^{t+1}\right)\right)}\right]. \tag{10}$$

TC > (<) 1 indicates that the best practice technology in the period $t+1$ is closer (further) to the biennial best practice frontier than the best practice technology in the period t. In other words, when TC is greater (smaller) than one, there is a progress (regress) of the technology.

In order to calculate and decompose the BML index, we need to specify four DDFs: two functions with observations under consideration and biennial production set—namely $\vec{D}^B\left(x^t, y^t, b^t\right)$ and $\vec{D}^B\left(x^{t+1}, y^{t+1}, b^{t+1}\right)$ —and two functions where observations and production set are from the time periods under observation—namely $\vec{D}^t(x^t, y^t, b^t)$ and $\vec{D}^{t+1}(x^{t+1}, y^{t+1}, b^{t+1})$. Following Chung et al. (1997) and Kumar (2006), in order to compare the BM and the BML, we consider the direction vector $g = (y, -b)$. Thus, we seek for the maximum proportional contraction of negative externalities and the maximum proportional expansion of positive outputs.

The contemporaneous DDF $\vec{D}^t(x^t, y^t, b^t; y^t, -b^t)$ for the kth airline can be computed as follows:

$$\vec{D}^t\left(x_k^t, y_k^t, b_k^t; y^t, -b^t\right) = \max\beta \tag{11}$$

$$\text{s.t.} \sum_{k=1}^{K} \lambda_k x_{\text{kn}}^t \le x_{0n}^t \quad (n = 1, \ldots, N), \tag{11a}$$

$$\sum_{k=1}^{K} \lambda_k y_{\text{km}}^t \ge y_{0m}^t + \beta y_{0m}^t \quad (m = 1, \ldots, M), \tag{11b}$$

$$\sum_{k=1}^{K} \lambda_k b_{\text{kj}}^t = b_{0j}^t - \beta b_{0j}^t \quad (j = 1, \ldots, J), \tag{11c}$$

$$\sum_{k=1}^{K} \lambda_k > 0, \tag{11d}$$

$$\lambda \ge 0. \tag{11e}$$

where the variables λ represent the targets identifying the linear combinations of the efficient DMUs. Given the similar size of the airlines, the model technology is specified as constant return to scale (constraint 11e). Notice that because of weak disposability on undesirable output (axiom 2e), the constraint (11d) appears as equality. The computation of $\vec{D}^{t+1}(x^{t+1}, y^{t+1}, b^{t+1}; y^{t+1}, -b^{t+1})$ is similar to Eq. (11) with the exception of the superscript t that becomes t+1 for all the variables.

The biennial DDF $\vec{D}^B(x^t, y^t, b^t; y^t, -b^t)$ for the kth airline can be calculated by solving the following DEA model:

$$\vec{D}^B(x_k^t, y_k^t, b_k^t; y^t, -b^t) = \max\beta \tag{12}$$

$$\text{s.t.} \sum_t^{t+1} \sum_{k=1}^{K} \lambda_k x_{\text{knt}} \le x_{0n}^t \quad (n = 1, \ldots, N),$$

$$\sum_t^{t+1} \sum_{k=1}^{K} \lambda_{\text{kt}} y_{\text{kmt}} \ge y_{0m}^t + \beta y_{0m}^t \quad (m = 1, \ldots, M),$$

$$\sum_t^{t+1} \sum_{k=1}^{K} \lambda_{\text{kt}} b_{\text{kjt}} = b_{0j}^t - \beta b_{0j}^t \quad (j = 1, \ldots, J),$$

$$\sum_t^{t+1} \sum_{k=1}^{K} \lambda_{\text{kt}} > 0,$$

$$\lambda \ge 0.$$

Model 12 is similar to the contemporaneous DEA. The only main difference is the double sum on the left side of the constraints highlighting the biennial nature of the PPS. Indeed, the benchmarks originate from the two time periods t and t+1. $\vec{D}^B(x^{t+1}, y^{t+1}, b^{t+1}; y^{t+1}, -b^{t+1})$ is computed according to Eq. (12) except for the fact that the superscript t becomes t+*1* for all the variables.

The biennial Malmquist (BM) index is computed as presented in Pastor et al. (2011). The main difference between BM and BML indices is that BM does not consider the production of bad externalities in the distances computation. We note that Oh (2010) describes the equivalence between the global Malmquist–Luenberger and the global Malmquist indices. Such equivalence is completely valid also for the BML and the BM index.

3.2 The Production Process

A technical production process consists of physical inputs transformed into physical outputs. In the airlines business, it is crucial to exploit the available capacity since operating flights with empty seats is extremely inefficient from an economic point of view. Hence, we benchmark airlines on the basis of their ability in maximizing capacity utilization. We measure the total capacity offered by airlines as the number of available seats kilometres (ASK) as a measure of capacity reserved to passenger service. Another input essential for the business is the labour. We therefore include the staff (i.e., pilots, co-pilots, and cabin crew) as a second input of the production process. These inputs utilization depend on airlines' strategy in terms of combined aircraft sizes, flight frequencies, and routes offered. Airlines can exploit this capacity by transporting passengers. Hence, the desirable output of the production process is the number of revenue passenger kilometres (RPK) the airlines performed. Finally, the CO_2 emissions represent the undesirable output of airlines' production process. This is a simple but effective representation of airlines' production process, which evaluates an efficient airline when carrying as many passengers as possible and producing as little CO_2 as possible given the flights operated (i.e., the inputs). Notice that, given the level of inputs, airlines can increase their efficiency (i) by reducing CO_2 emissions hence operating greener aircrafts and/or (ii) by increasing passenger volumes.

3.3 Data Collection

Our airline sample includes the 12 largest airlines over the 2010–2017 period in terms of total passengers carried. We excluded from the analysis carriers not operating in each single year (e.g., because of failure/mergers), as well as low-cost carriers (e.g., Southwest, Ryanair, etc.) and purely cargo carriers (e.g., Cargolux, DHL, etc.) due to their significantly contrasting business models. The resulting sample covers around one bn revenue passengers in 2017. Table 1 enlists the carriers included in the analysis.

For each flight operated by the abovementioned airlines during the analysed period, we collected data regarding the following variables:

Table 1 Worldwide airlines included in the dataset

IATA code	*Name*
AC	Air Canada
CA	Air China
NH	All Nippon Airways
AA	American Airlines
BA	British Airways
MU	China Eastern Airlines
CZ	China Southern Airlines
DL	Delta Air Lines
EK	Emirates
LH	Lufthansa German Airlines
TK	Turkish Airlines
UA	United Airlines

- passengers;
- available seats (AS);
- available seats kilometres (ASK);
- frequency;
- flight length; and
- aircraft model.

The data source is represented by the Official Airline Guide (OAG) database. By aggregating the OAG flight level data, we obtain yearly values for AS, ASK, frequency, and kilometres flown for each airline. Furthermore, we collected the yearly data from the ICAO data+ for each operating airline considered in the analysis:

- revenue passenger kilometres (RPK, millions), that are the product between the total number of revenue passengers and the corresponding flown distances;
- total employees in full-time equivalent.

All the variables are collected with respect to the code of the operating airline. In order to compute the CO_2 emissions, we adopt the procedure developed in Scotti and Volta (2015):

1. We group all the operating flights included in OAG database by (i) airline, (ii) aircraft model, and (iii) distance flown. By considering operating flights only, we can exclude non-operating duplicate flights arising from codeshare agreements.

2. We use the Eurocontrol emission calculator to compute the CO_2 produced for each flight of an aircraft model by the specific airline.
3. We aggregate the CO_2 produced at a flight level in order to obtain the total amount of CO_2 yearly emitted by each airline.

4 Data Trends

Figure 2 shows the trends in terms of CO_2 emitted. When considering our airline sample, the emissions of CO_2 increased around 45% with respect to the 2010 level. The increase may be explained by the growth in term of passengers (+42%) and movements (+25%). Indeed, coupling emission levels with passengers carried (CO_2/pass), revenue passenger kilometres (CO_2/RPK), and available seats kilometres (CO_2/ASK), we observe an average decrease: −2, −7, and −4%, respectively. It is important to notice that the decrease in the relative emissions may be due to the less than proportional increase in movements (+25%) and the relative increase of the average stage length. Indeed, the considered airlines

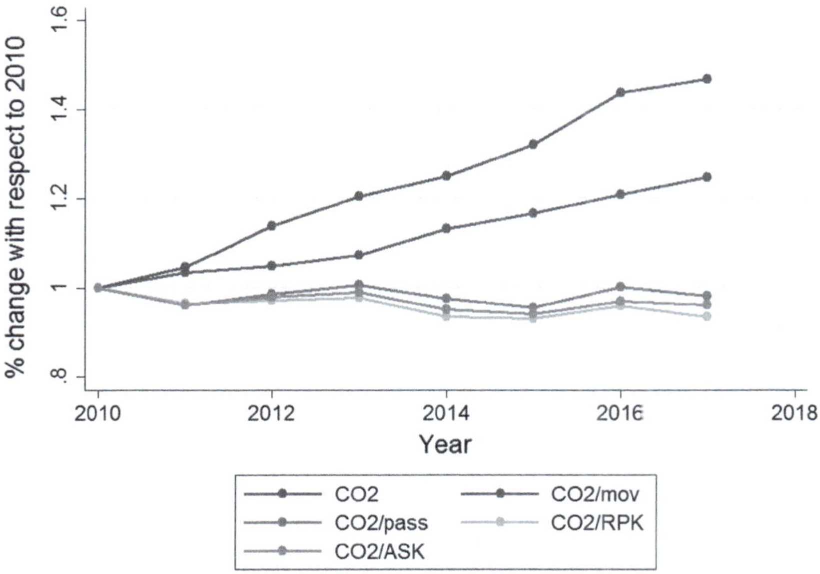

Fig. 2 CO_2 emissions evolution

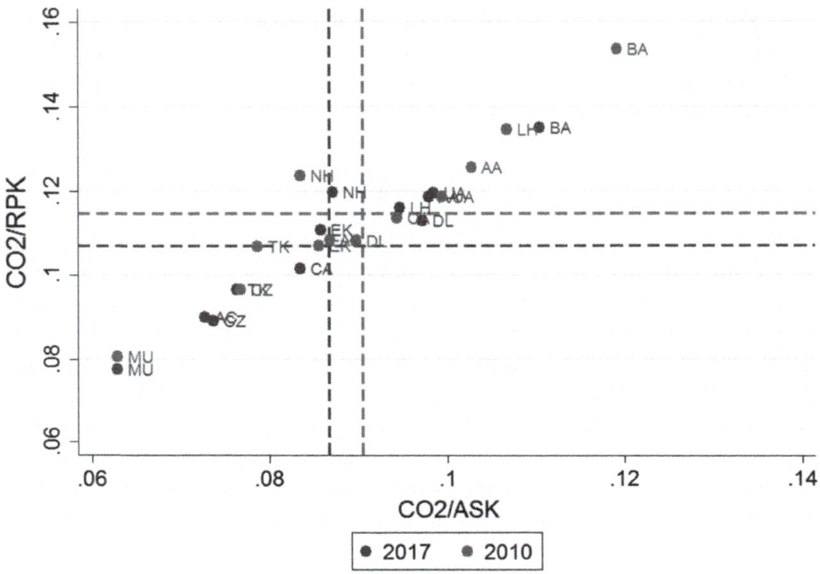

Fig. 3 Average levels of CO_2/ASK and CO_2/RPK (2000 and 2010)

increased the average stage length from the ~1.500 km per flight in 2010 to the ~1.800 km per flight in 2017.

Similar improvements can be depicted in Fig. 3, which shows the amount of CO_2 kilograms per available seats kilometres (ASK) and per revenue passenger kilometre (RPK) of each airline considered in our sample for both 2010 and 2017. The CO_2/ASK is a value strongly influenced by the aircraft model owned by the airline (i.e., greener aircraft means, ceteris paribus, a lower level of CO_2 emissions), but also by the network offered. In fact, longer flights may result, ceteris paribus, in lower CO_2/ASK values given that the CO_2 produced during the landing and take-off cycle is divided by a higher number of kilometres. The CO_2/RPK is influenced also by the airlines' capability of filling the seats offered. In this sense, a lower distance between the two values is a signal of higher average airline's load factor. Figure 3 does not consider the CO_2 produced by cargo flights in order to fairly compare airlines. The dotted lines represent the average values (red for year 2010, blue for 2000). The general shift of the observations towards the bottom-left side of the picture reveals a decrease over time in both CO_2/ASK and CO_2/RPK. Average values decrease in ten years of around 7% (from 0.115 to 0.107 of CO_2/RPK) and of 4% (from 0.091 to 0.086 of CO_2/ASK). Nevertheless, not all the airlines experienced the evolution described. For example, Delta Air Lines and Emirates show a (small) increase in both the indicators while All Nippon Airways (NH) and United Airlines (UA) show an increase in CO_2/ASK and CO_2/ASK, respectively.

The decrease in CO_2/ASK may be explained by a shift to more environmentally friendly fleets. In this regards, Fig. 4a, b show the most used aircraft in terms of movements and distance flown comparing the 2010 and 2017 values. The Canadair CRJ Series 705 was the most operated aircraft model in terms of both movements and distance flown in 2010. This is likely due to the high level of utilization of this model in North America. The CRJ7 has on average 75 seats with a maximum flying range of around 3.500 kilometres. In contrast, 2017 is characterized by a higher utilization of the Airbus A320 and the Boeing 737-800, which are the highest in terms of movements and in the top 3 with respect to the kilometres flown. The B738 has on average 150 seats (depending on airline configuration choices) with an average range of 5.000 kilometres, and equally the A320 (the Airbus antagonist) has similar specifications in terms of seats and range. Figure 4a, b highlight a renewal in the fleets, old aircraft (e.g., McDonald Douglas—MD81, Boeing 757-200—B752) have been substituted by newer models, such as the Boeing 787 Dreamliner (B788).

When analysing passengers transported, the last few years have shown an increase of in-air transport activity, mostly as a result of the ongoing recovery from the global economic downturn caused by the financial crisis in 2009. Figure 5 shows the evolution of the total passengers transported, the total movements performed, and the total kilometres flown over this time period. Values are considered as the annual average evolution with respect to 2010 over our airlines sample. The number of passengers transported increased around +42%, while movements increased with a lower rate (+25%). Similarly, the total amount of kilometres flown increased by 41% over the same period (see Fig. 5).

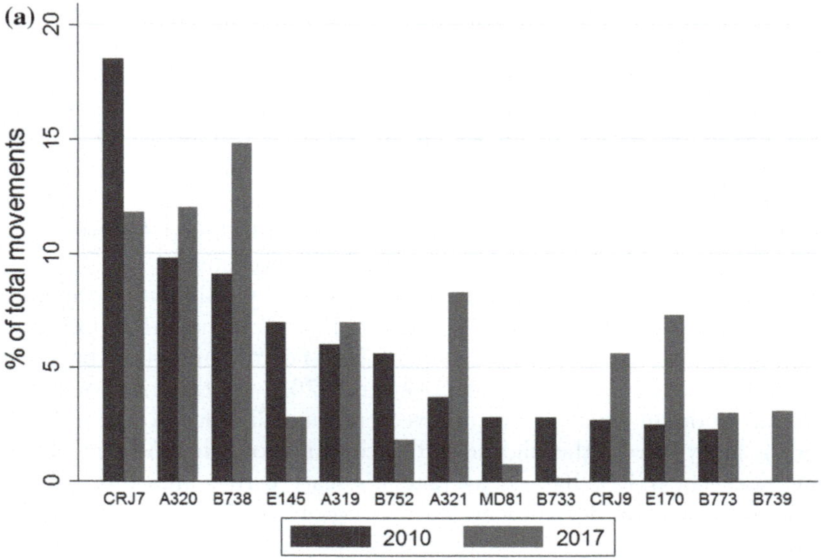

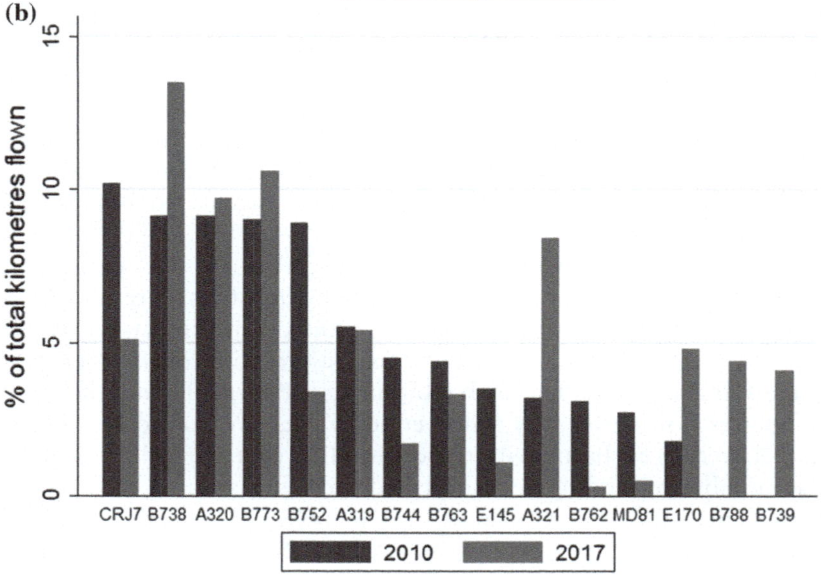

Fig. 4 2010 and 2017 % of aircraft movements and kilometres flown by aircraft model

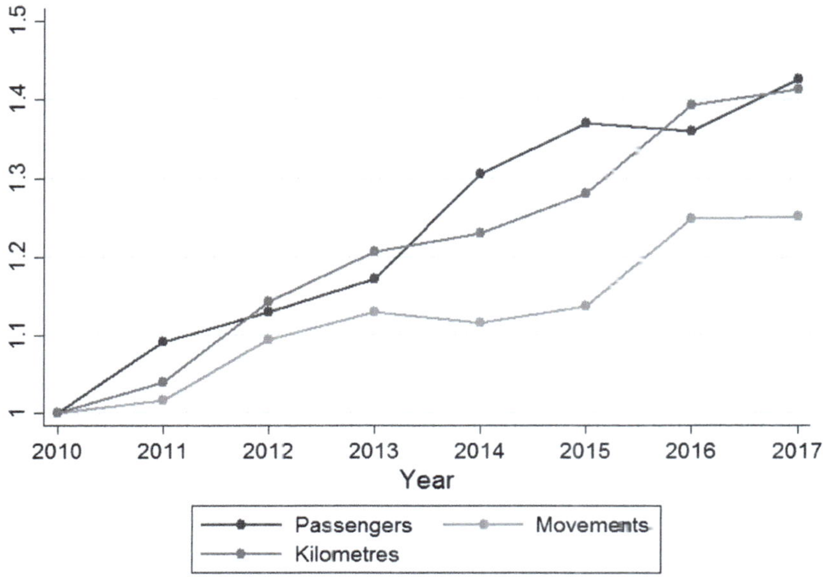

Fig. 5 Evolution of passengers, movements, and kilometres flown since 2010

It is important to notice that the difference in growth levels may have different origins. The first cause is the increase in load factor (i.e., the number of revenue passenger with respect to the available seat). Indeed, the average load factor recorded an increase from the 82% of 2010 to the 84% of 2017. The increased size of the aircraft operated may be another important source for the difference. Figure 6 shows an increase in the average aircraft size in terms of seats per movement (from 131 to 156; +21%). This data reflects the fleet switch emphasized in the previous pictures. This trend may partially reflect a reaction to higher airport congestion and/or to increased slot values. These factors may indeed force airlines to adopt fleets composed by bigger aircraft. However, other practices may influence the per flight average available seats, such as an increase of long-haul flights (as shown in Fig. 6 by the increase in the average stage length) or a change in the seat configuration (e.g., reducing first class/business seats in favour of dense economy seats).

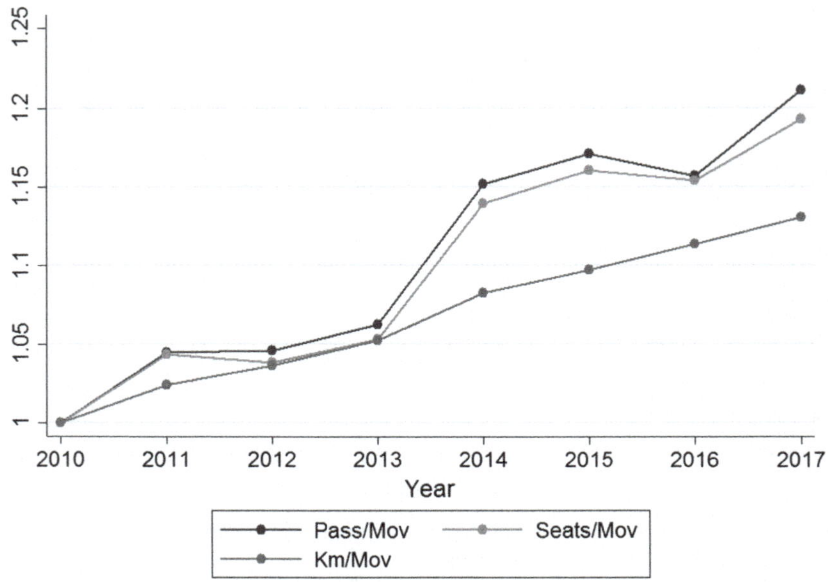

Fig. 6 Air traffic evolution since 2010

5 Results and Discussion

The improvements shown in the previous section can be loosely referred as airlines productivity. However, the univariate analysis of the indicators may miss the complex interrelations existing between the factors. In order to capture both the technical and the efficiency improvements over time, it is necessary to evaluate productive performances with proper instruments, such as the indices presented in the methodology section. Therefore, we present the results obtained from the best practice biennial technical frontier without considering the environmental dimension (BM) and considering the environmental dimension (BML).

Figure 7 shows the BM and BML smoothed values for the airline sample considered. Values are the estimated productivity change over biennial periods. Generally, both indices are above the unit showing an increase of airlines productivity. The first biennial period of analysis is the only one showing a regress of productivity, probably due to the recovery efforts related to the financial crisis. When considering the differences

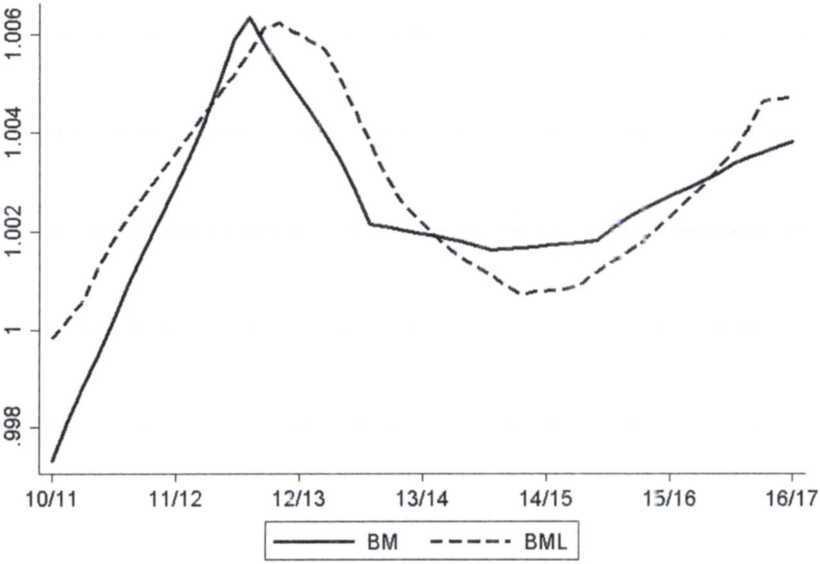

Fig. 7 BM and BML smoothed values

between the two indices, it is not possible to clearly identify a diversity as both indices follow a very similar trend. However, the overall average rate of the BM productivity index shows a lower improvement (+0.19%) compared to the BML productivity index (+0.25%). This comparison indicates that the airlines are on average improving their productivity performances while reducing at a faster pace the emission produced.

The productivity improvements can be deconstructed in two main sources: technical and efficiency changes. The former indicates a shift of the production frontier (i.e., the industry is moving towards a lower utilization of inputs and a higher production of outputs), and the latter describes the ability of the single airlines to catch up with the best production frontier. Figure 8a, b show the smoothed decomposition of the BM and BML indexes for our airline sample in efficiency change and technical change. As for the productivity indices, values above the unit indicate a catching up for the efficiency change index and a progress for the technical change. Starting with the estimated efficiency change (Fig. 8a), both the BM and BML indices show a similar trend starting

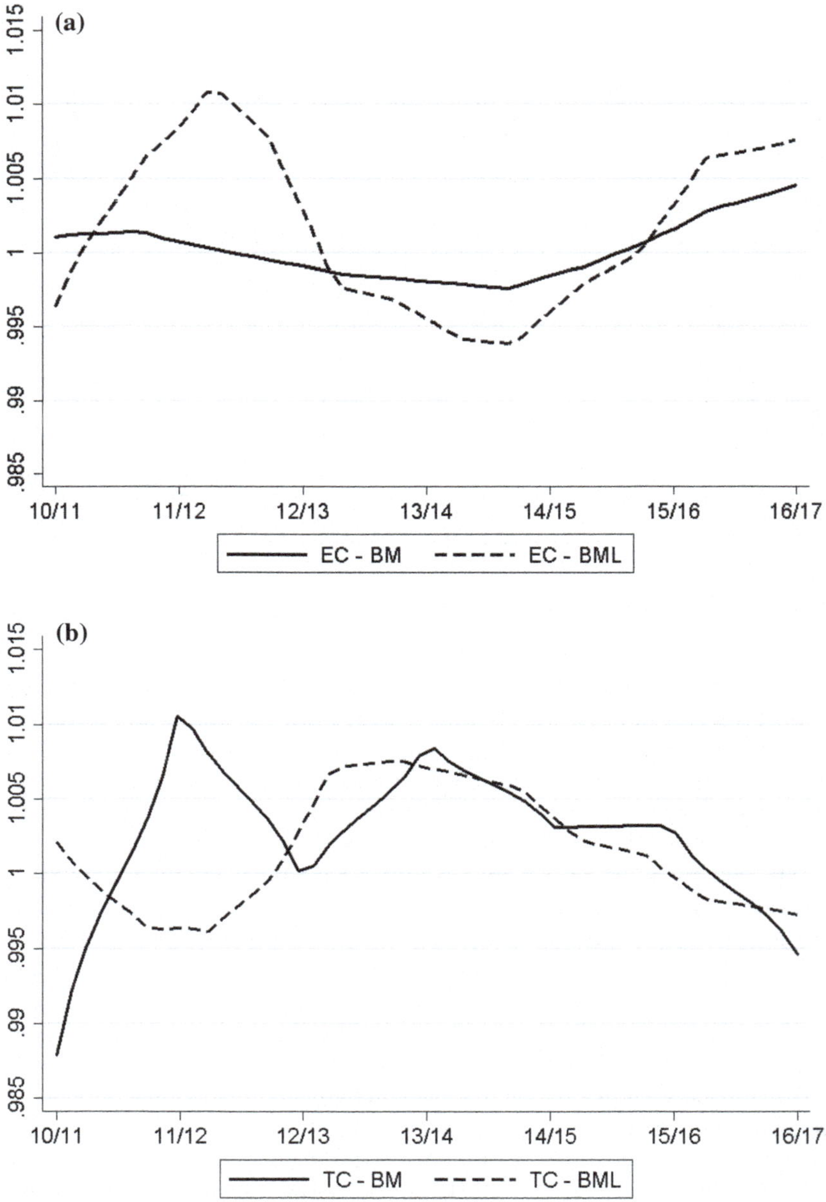

Fig. 8 Smoothed efficiency change (EC) component and smoothed technical change (TC) component

from 2012 while behaving differently in the first years of analysis. More specifically, the index considering environmental efficiency (BML) presents a big catching up effect between the periods 2011–2013. This may be the effect of the recovery from the financial crisis when inefficient airlines were able to increase the passenger transported while emitting an equal amount of CO_2 (i.e., improved load factors). However, it is important to notice that both BM and BML efficiency changes are rather low with average catching up effects of +0.08 and +0.12%, respectively. When analysing the technical change (Fig. 8b), both indices are following a similar trend with, again, the only exception for the period 2011–2012. The figure shows a positive technical change: Airlines production process has moved towards a more efficient utilization of inputs and a higher production of outputs. Other than better managerial ability in increasing load factors, this may be due as well to the switch to more modern aircraft, as highlighted in the previous Fig. 4a, b. Technical change has been the major factor of productivity growth; the overall biennial averages of technical progress are +0.11% (BM) and +0.15% (BML)

6 Conclusions

Due to the increasing amount of residential development surrounding airports and the concerns regarding the climate change, the continued growth of commercial air transport and the relative increase in air pollution have become significant concerns for local governments and international authorities. Based on finite planetary resources, aviation cannot be considered sustainable in the long term, but it is realistic to reason and plan policies that have an impact in the medium term (ensuring safety, efficient optimization of available capacity, optimization of the balance between social, economic, and environmental imperatives, minimizing negative impacts, and using more efficient ATM resources, equipment, and technologies). Clearly, environmental aspirations must be balanced with the commercial necessities of maintaining a commercially viable air transport industry. Aviation is a relevant driver of economic growth, employment, trade, and mobility for regional development. For these reasons, it is imperative to ensure high standards of air safety and security, to support innovation, and to reduce its environmental impact.

In this chapter, we considered a modelling approach to targeting productivity and environmental impact in a unique and correct measure of airlines development.

We implemented a BML index to consider CO_2 emissions in airlines' productivity assessment. By including negative externalities, our modelling approach estimates a comprehensive measure of productivity, technical, and efficiency change. We compared an environmental-sensitive index (BML) and a traditional index (BM) that considered a sample of 12 major worldwide airlines for the 2010–2017 period.

The airlines' production process is analysed on the basis of two input variables: the number of available seats kilometres (ASK), describing carriers' capacity in terms of passengers and airlines staff (i.e., pilots, co-pilots, and cabin crew). Two output variables are considered: one desirable describing the traffic levels (i.e., passenger volumes) and one undesirable describing the environmental impact in terms of CO_2 emissions.

Observing the BM and BML productivity index (+0.19 and +0.25% respectively), we state that the airlines are, on average, improving their productivity performances while reducing at a faster pace the emissions produced. Our analysis shows how airlines' production process has moved towards a more efficient utilization of inputs and a higher production of outputs (positive technical change). Airlines' productivity is improving over the recent period, with technical change being the major component of such increase. The improvement in fleets helped airlines to reach higher economic and environmental performances.

While our results are encouraging, question marks remain as to the extent to which incremental improvements in productivity alone can be expected to secure an environmentally sustainable future for air transport. Certainly, significant challenges remain in this regard, not least in terms of the projected future growth in air traffic in the near-to-medium term. Furthermore, it is worth noting that our analysis excluded both low-cost and cargo carriers, with each playing a notable role in the global picture. Consequently, it is important that airline productivity gains are successfully coupled with, and driven by, the creation of suitable, equitable, and robust environmental policy measures on a global scale, implemented in a climate of operational and technological innovation. Arguably, it is under these conditions that the potential for securing a more sustainable future for air transport is most realistic.

References

Adler, N., & Volta, N. (2016). Accounting for externalities and disposability: A directional economic environmental distance function. *European Journal of Operational Research, 250*(1), 314–327.

Budd, T. (2017). Environmental impacts and mitigation. In L. Budd & S. Ison (Eds.), *Air transport management* (pp. 283–306). Surrey, UK: Ashgate.

Chung, Y. H., Färe, R., & Grosskopf, S. (1997). Productivity and undesirable outputs: A directional distance function approach. *Journal of Environmental Management, 51*(3), 229–240.

Färe, R., Grosskopf, S., & Pasurka, C. A., Jr. (2007). Environmental production functions and environmental directional distance functions. *Energy, 32*(7), 1055–1066.

ICAO. (2016). *On board, a sustainable future: Environmental report 2016* [Online]. Available at: https://www.icao.int/environmental-protection/Documents/ICAO%20Environmental%20Report%202016.pdf. Accessed 31 July 2018.

Kumar, S. (2006). Environmentally sensitive productivity growth: A global analysis using Malmquist-Luenberger index. *Ecological Economics, 56*(2), 280–293.

Landrum and Brown, Inc., Environmental Consulting Group, Inc., Primera Engineers, Ltd., & Muller and Muller, Ltd. (2012). *Guidebook for incorporating sustainability into traditional airport projects.* Airport Cooperative Research Program: Report 80, Transportation Research Board for the National Academies, Washington, DC.

Lee, D. S., Fahey, D. W., Forster, P. M., Newton, P. J., Wit., R. C. N., Lim, L. L., et al. (2009). Aviation and global climate change in the 21st century. *Atmospheric Environment, 43*(22–23), 3520–3537.

Oh, D. (2010). A global Malmquist-Luenberger productivity index. *Economics Letters, 88*(2), 266–271.

Pastor, J. T., Asmild, M., & Lovell, C. A. K. (2011). The biennial Malmquist productivity change index. *Socio-Economic Planning Sciences, 45*(1), 10–15.

Scotti, D., & Volta, N. (2015). An empirical assessment of the CO_2-sensitive productivity of European airlines from 2000 to 2010. *Transportation Research Part D, 37,* 137–149.

Suau-Sanchez, P., Pallares-Barbara, M., & Paul, V. (2011). Incorporating annoyance in airport environmental policy: Noise, societal response and community participation. *Journal of Transport Geography, 19*(2), 275–284.

CHAPTER 5

The Analysis of the Sustainability Commitment Formulation and Implementation for the Selected Airlines

Margaryta Radomska and Larysa Cherniak

1 Introduction

The transportation industry, including automobile, water, rail and air transportation, is one of the major air polluters of the urban areas. It also gives origin to the pollution of surface water bodies, the formation of solid waste and the creation of noise pollution and electro-magnetic fields.

The intensity of impact, emissions and other forms of pollution produced by cars exceed the corresponding magnitude of environmental impacts compared to other types of transportation. However, the impact significance from cars is almost equal with the impact from air transportation at its greatest intensity, as air transportation is able to affect

M. Radomska (✉) · L. Cherniak
Department of Ecology, National Aviation University, Kyiv, Ukraine

L. Cherniak
e-mail: specially@ukr.net

T. Walker et al. (eds.), *Sustainable Aviation*,
https://doi.org/10.1007/978-3-030-28661-3_5

global natural processes. As a result, a target for national governments and international organizations is to control and regulate the changes of the environmental quality and its functionality under the influence of aviation. The widely accepted concept of sustainable development promotes the creation of corresponding legal frameworks, and it forces enterprises to declare their willingness to change their industrial practices toward environmental efficiency. The aim of the current research is to evaluate the modern perception of air transport environmental impacts by the operators of the market—airlines.

2 Environmental Management Systems for Transport Facilities

The process of solving any socioeconomic problem includes the application of a series of consistent measures based on managerial decisions. However, the efficiency of such activities depends directly on identifying the problem and on clearly formulating the goals for tackling each problem. According to the generally accepted position in political and law sciences, the goals are defined in the framework of policy formation in one or another field. The above refers to the policy in the environmental sphere as a whole and in its directions in particular. The goals and objectives of such policy are dynamic, which must ensure the timely resolution of existing problems and respond to new challenges.

The contradictions between the economy and the environment are solved in the process of scientific and social development by regulating the structure, organization, technology and methods of economic activity in terms of their compliance with natural processes (Čančer 2000). The need to regulate the relationship between economy and environment arises from two opposing but interacting phenomena; due to modern production trends and the market economy, enterprises are forced to modernize their production in order to achieve the necessary rates of economic development, growth of production volumes and profits, under the pressure of severe competition at external and internal markets. The expansion of production leads to growing environmental pollution and natural resources consumption. In this regard, ecological management with its environmental management tools should become the art of making efficient management decisions in order to improve the environmental performance of the enterprise without compromising its economic perspectives.

Two decades ago, the widespread opinion was that an enterprise investing in nature protection was a no-win proposal for managers of any industry (Walley and Whitehead 1994). Today being green is no longer considered a cost of doing business. Due to risen public awareness about the real intensity of their environmental impacts, preventing any threat to the environment has become a strategic objective for industries. The general public's gained knowledge has affected the buying habits and consumption trends, and thus, it has affected the revenue of these enterprises Under such an influential driver, companies have started to join the "green" movement. In this new world, both business and the environment can win. Being green is no longer a cost of doing business; it is a catalyst for innovation, new market opportunity and wealth creation (Clarke et al. 1994).

The idea that demonstrating interest and reducing corporate environmental impact can bring increased profitability for business is widespread now, as shown in studies by various authors in the mid-1990s (Bennis et al. 1995; O'Callaghan 1996; Reijnders 1996; Ruffing 1998; Ulhoi et al. 1996; Brandrup 2005). If producers change the design of a product to cut the input of valuable natural resources, consumption of energy or output of pollutants to the environment, it has a direct influence on future profits due to the reduced self-cost of production, as it was shown on the examples of number of international companies, including 3M and the Dow Chemical Company (DeSimone and Popoff 1997), local authorities and small- and medium-sized companies (Welford 1994) and educational establishments (Winter 1995). But for these, considerations of nature protection must be reflected and implemented in all elements and stages of production process. So, there must be an integral system of actions interwoven in the everyday work of an enterprise. This comprehensive approach is provided by the procedures and instruments of environmental management.

Environmental management is a special management system that is based on a regulatory process aimed at preserving the quality of the environment and ensuring the standard quality of the social, environmental and economic media.

With reference to the air transport industry, environmental management would include solutions for natural resource use optimization, for both economic efficiency and nature protection points, organizational structure that meets green ideals; environmental marketing and human resources management; environmental and corporate culture, behavioral patterns; and interaction with the authorities and the public.

The functions of environmental management include common tasks (formulation of goals, forecasting, planning, organization, decision making, etc.) and special ones, such as technical tasks (creation and implementation of low-waste technologies, etc.), environmental accounting, choosing performance indicators and formulating evaluation scales for them. All environmental management functions are closely related to each other and to all the components of enterprise activity. This is especially important for the transportation sector as its diversified impacts on the environment are hard to control and mitigate.

The newest trend is to perceive the environment as one of the forms of natural capital, in a way similar to material fixed assets (Collados and Duane 1999). Therefore, damage to the environment embodies financial damage and losses of assets, which are often non-renewable or need huge investments for restoration.

Within the framework of the nature management for transport enterprises, there are three general types of capital involved in the process of services production and delivery: anthropogenic (artificially created) capital, like buildings (terminals, hangars, storages, warehouses, fuel depots, access roads, runway, etc.), which can be increased or reduced by our decision; critical natural capital (the ozone layer, the global climate, biodiversity, untouched areas, Antarctica, etc.) includes the necessary natural resources for life that cannot be replaced by anthropogenic capital; and other natural capital, which includes renewable natural resources and some finite mineral resources (these are various energy carriers, both conventional and alternative) that can be fully or partially restored or replaced due to modern achievements of science and technology.

Thus, the need to create an efficient system of environmental management is obvious in terms of volumes of natural resources, consumed by the enterprises of any transport industry, without production of any visible product except services for the passengers.

The first step toward creating an environmental management system at an enterprise is a clear awareness about the presence of environmental aspects in its activities. Having defined the most important aspects, one can set the goals and blueprint their achievement, that is, starting to develop and implement the environmental management system. The formulated goals and a detailed comprehensive plan are given in the form of a strategy, which is promulgated as an environmental policy.

It's not compulsory to have an environmental policy, but it is necessary if an enterprise is planning to implement an environmental management standard, such as the European Union Eco-Management and Audit Scheme (EMAS), BS 8555 or ISO 14001. It can also help develop the positive external image of the company by demonstrating commitment to reduce environmental impacts to customers, potential clients, investors, suppliers and business partners. This in turn can lead to financial benefits, such as increased investment, customer sales and market share.

However, many transport enterprises are old and conventional; therefore, they practically have no environmental principles in the general management system, and many of them have no possibility to do that due to political, economic or social reasons. Airlines are not old nor conventional, and they do not have limited resources, with exception of those that operate in developing countries at a domestic level. Their willingness to reduce profit in the name of common good is not higher than that of any other enterprise, and this is the most important drawback of environmental management. This reasoning puts under question the efficiency of the global improvement of industry's environmental performance; environmental management system is implemented on a voluntary basis, on the initiative of the entity itself. Of course, declaring its own environmental policy and implementing the system of environmental management, the enterprise eliminates the formalism of administrative regulation, but at the same time, it can postpone and modify the provisions of the plan instead of exerting to achieve the best.

Thus, implementation of environmental management will be considered economically useful and expedient by airlines if the following factors are present:

- significant reduction in production costs and resources consumed, seized raw materials, water, energy and penalties for pollution;
- improvement of product quality and as a result increased competitiveness;
- improvement of relations with state authorities or opening access to certain types of state support;
- expansion of markets and attraction of new consumers (for airlines in developed countries access to new markets is impossible without compliance of environmental standards);
- new level of technological development and innovation, which can bring all the above-mentioned benefits and more by providing qualitatively new services.

A quick look at this industry's situation shows that airlines are paying increased attention to their activities' environmental effects, presenting environmental policies and strategies. But a more thorough approach to the manifested documents can reveal to some extent a different picture. The task of the following research is to assess the quality and content of universal air transportation companies' available environmental policies, define their drawbacks and develop recommendations for their improvement.

3 The Rating of Air Transportation Impacts on the Environment

As it has been previously said, efficient environmental management must start with identifying the environmental problems correlation with the transport sector activity, as well as defining the natural resources needed and affected by the industry's activities.

In airline sector, the problem lays in having an obsolete picture of how the air transportation service affects the environment. Most of the practitioners and customers believe that aircraft sole environmental impact are emissions of fuel combustion products and noise pollution generated by jet engines. But these represent only the beginning of a long list of impacts.

Air transportation is an industry that has some direct and clear, but also some subtle and less obvious, impacts on vital resources. This raises attention to its activity. Flights are important contributors to greenhouse emissions, global warming and global dimming due to the contrails left after the plains.

The effect of aircraft emissions on the thermodynamic properties of the atmosphere is not as simple as it seems. From one side, GHG emissions will contribute to global warming, if released in the troposphere, but the effect is unclear at the stratosphere level. Here, ozone produces the opposite, i.e., a cooling effect, and water vapor induces its impact. The contrails and resulted cirrus clouds are also able to mitigate expected temperature rise by dimming the atmosphere. This results in raising nighttime temperatures and/or lowering daytime temperature, which was proved by observations during the three days following the 11 September 2001 attacks (Travis et al. 2002). Emissions of particulates, including sulfate aerosols in the atmosphere, also produce opposite effect by reducing radiating potential and causing dimming effect. As these are the impacts on a global scale, they should be given the first rank in the rating of potential aviation impacts and it must be obligatorily mentioned in the environmental policies. However, the possible approaches to the problem should not only cover CO_2 emissions reduction, if not, it should cover all the complex climate effects. Moreover, recent studies show that there could be a problem at the level of microclimate local to the airfield (Cherniak and Radomska 2017). Microclimate problems put communities under threat, and therefore, air transportation practices conflict with the basic principle of sustainable development and the environmental management system—support and protection of local communities.

Gases released by the engines of proposed high altitude supersonic aircraft on the ozone layer have the potential to destroy significant quantities of ozone in the stratosphere, depending on the altitude of release. However, the depletion of ozone could be substantially reduced due to conversion of nitrogen oxides into nitric acid on sulfate aerosol particles. The applied modeling also suggests that significant ozone depletion could result from the formation of additional polar stratospheric clouds produced by the injection of H_2O and HNO_3 from the aircraft engines (Tie et al. 1996). This effect of aviation is of course a global threat, but it is hard to manage due to high uncertainty and therefore cannot be granted the first level of importance for environmental management strategies.

The impact of ground emissions is of the second priority due to significance of it in the environmental quality of local communities. This is also a very manageable impact as it could be tackled with a combination of organizational and technical solutions of low complexity. Noise pollution lays in the same environmental level as ground emissions because of its environmental disturbance and manageability.

The next group of impacts, soil and underground waters pollution as a result of fuel spills and leakages, as well as the application of anti-acing and de-icing agents, is not obvious to most passengers and even operators of air transportation. Of course, these factors have seasonal or geographical character. They also depend on the level of technological culture and quality of the equipment and installations. But it is a good field for improvement when moving in the direction toward sustainable operations.

The contribution to habitat fragmentation and reduction of local species diversity and population is also a relevant problem for airports operated by airlines. The impact is formed due to aggregation of the above-mentioned factors of influence, and the systems of biological safety applied at the airports.

The energy and resource efficiencies are among the top initiatives targeted by all industries, including air transportation. It gives possibility to reduce consumption of natural capital, both at the cost of direct decrease of resources used and at their degradation due to pollution. Efficiency improvement usually gives noticeable economic benefits and therefore is treated favorably by managers and economists of a company. However, the equal sign should not be placed between energy efficiency and environmental safety, as environmental safety is a concept of higher level and importance, and must be a supreme goal, while energy efficiency is one of the ways to get there. Another way is of course the introduction of alternative energy sources and power generating capacities, which are a change of essence and not only the transformation of the existing practice.

There is, of course, a range of environmental issues due to aviation activities, which cannot be tackled under the modern level of technological development: electro-magnetic pollution from radar and other types of equipment. Likewise, it is impossible to provide "zero emission" and "zero waste" technologies for air transportation. However, the strategy of environmental performance improvement must be formulated in a way that invests efforts in the solution of real problems of the highest priority. So, the following analysis is aimed at the comparative study of the airlines' manifested environmental policies, defining their relevance to the reality of the industry.

4 Method of Analysis

Environmental policy is a system of activities related to the influence of society on nature. Environmental policy is a set of intentions and principles defined by the organization concerning the environmental performance of its activities, which creates the basis for the development of specific goals and objectives. At the same time, the environmental strategy is defined as a set of specific goals and objectives, based on real possibilities and terms of their achievement, in the sphere of environmental protection and the use of natural resources.

The environmental policy could be developed at different levels with various levels of enforcement. In case of transport sector and civil aviation, the following levels are applicable:

- global: international, political and economic initiatives in the form of declarations and protocols opened to any country, ready to implement the provisions, goals and restrictions of the document—in the studied case, it is the ICAO or IPCC legal frameworks;
- state: social-economic policy, developed at the national level, built on the understanding of the effects and shortcomings typical for air transportation industry within a country. This is represented by a national strategy or plan;
- regional environmental policy: an administrative area policy, aimed at development and support of the air transportation sector with respect to the need to prevent the equal development and propagation of environmental problems;
- local environmental policy: normally developed by local authorities to control and limit the impacts on the environment at the territories adjoined to the airports and maintenance facilities;
- corporate environmental policy: the most realistic strategy as it presents the direct list of actions to be performed to achieve sustainability of operations.

Thus, the last position is of primary interest for the survey, especially its application by airlines. Airlines were chosen to be the object of research as they represent the most dynamic part of the industry; they possess ample opportunities in implementation of organizational and technological solutions, as well as financial and labor resources. At the same time, airlines are basically the operators of the sources of the impacts—aircrafts and airports with their facilities.

In the process of analysis, the following criteria were identified as the most important for the overall evaluation:

1. Clear formulation of targets and goals set by the company;
2. Determined deadlines for the implementation of the goals or plan of their achievement in the form of steps with specified subtargets;
3. Correspondence to the list of priority problems for aviation industry;
4. Obligatory control over performance of the plan and strategy overall;
5. Specified indicators of success with numerical equivalents, if applicable;
6. Reporting procedure: periodicity of results presented to the public.

For those airlines operating on domestic market, a well-developed environmental strategy should be a list of simple targets, which could be implemented under the current conditions or a set of indicators, reflecting environmental performance, regardless of type, size, location, etc., rather than a text with broad perspectives.

The selected set of environmental performance indicators, included in the environmental policy, must meet the following requirements:

- Comply with the developed environmental strategy of an enterprise and industry as a whole;
- Be simple and understandable;
- Enable operative analysis and adjustment in case of noticeable alterations of the external environment (natural, economic, social or technogenic).

The indicators can be expressed in absolute or specific values, and it is recommended to use the form that is more representative and comparative. If the common public is considered as a target audience, then various specific indicators are better suited. For example, in assessing achievements such as the reduction of emissions, energy consumption or materials consumption, quantitative indicators should be used to deliver information, but relative values could be more readily interpreted by non-specialists. Other useful quantitative indicators are financial parameters, suitable for both professionals and customers. As a result, the basis for environmental policy would be a plan to improve the values of environmental indicators, which were identified as important for the enterprise.

An important factor for the successful improvement of airlines activities' environmental performance is the above-mentioned openness and publicity of their environmental policy. Their environmental policy must be known and understood by the staff, partners and any individual that has an interest in the environmental aspects of the company's activities, its products and services, as well as the local communities, which undergo the effects of the enterprise activity. It will bring benefits to both society and the company. First of all, it has an advertising potential, which should be used for promoting the company's services on the market. Second, if the intentions of the airlines are given to the world, these will create noticeable, but bearable, press on the company by obligating it to implement the taken commitments and to spread the results over mass media for better feedback from the interested parties, including authorities.

The compulsory reporting also contributes to further development of the environmental policy and environmental management system with gradual approximation to the goal, substitution of the goals with more realistic ones and real-time corrections of their achievement.

For the purpose of comparative assessment, each of the studied policies was awarded a score made of the points for each of the above-mentioned criteria performed in the documented environmental strategy and the overall evaluation was made accounting additional factors, not covered by the criteria.

5 Analysis of Environmental Policies of the Selected European and World Airlines in Europe

We have randomly collected environmental policies of European and Universal Airlines. The study was based on the documents available online, because free and open to public access to the environmental policy is one of the essential attributes for its success. Generally, we have analyzed over 60 documents developed by companies at various levels, and here, we present the comparative analysis of the following selected airlines with different position on the market:

- *International airlines*
  - Air Arabia—major international (4 points),
  - Air Canada—major international (5 points),
  - Air France—major international (5 points),
  - American Airlines—major international (5 points),
  - British Airways—major international (6 points),
  - Delta Airlines—major international (4 points),
  - Emirates Airlines—major international (4 points),
  - Japan Airlines—major international (5 points),
  - Lufthansa—major international (6 points),
  - Turkish Airlines—major international (4 points),
  - United Airlines (USA)—major international (4 points),
  - Aeroflot—international (4 points),
  - Air Canada Jazz—international (4 points),
  - LOT Polish Airlines—international (3 points),
  - Thomas Cook Airlines (UK)—international (3 points),

  - Scandinavian Airlines—international (6 points),
  - Transat (USA)—international (4 points),
  - WestJet (Canada)—international (4 points),
  - Air India—international (4 points),
  - Air Berlin—international (currently not operating) (6 points),
  - Air Astana—international (3 points),
  - International Airlines of Ukraine—international (4 points).

- *National airlines* (with international segment, which is not dominant)
  - Australian Airlines—national (4 points),
  - Ethiopian Airlines—national (3 points),
  - Icelandair—national (3 points),
  - Irish Airlines—national (3 points),
  - Malaysia Airlines—national (4 points),
  - Norwegian Airlines—national (4 points),
  - Qatar Airlines—national (4 points),
  - Singapore Airlines—national (5 points),
  - South African Airways—national (4 points),

- *Local airlines* (with single international flights)
  - Alaska Airlines (USA)—local (3 points),
  - Maryland Airlines (USA)—local (3 points),
  - Southwest Airlines (USA)—local (3 points),
  - Qantas (Australia)—local (3 points).

As it can be extracted from the score set for each airline according to the established criteria, the typically failed criteria are the commitments to regular reporting with strict deadlines and numerical presentation of the targets. Another important weakness of the studied policies is underestimating the environmental impacts, which are objectively considered as the highest priority. It is also important to note that the best policies are not compulsorily presented by the major airlines. The formulation and presentation of environmental policies are very different around the world, as there are no strict regulations and recommendations about them. The lack of consistency due to the lenient regulations is seen even in the naming of the documents, which is set as "Environmental Policy" for most of the major airlines, including Delta, United Airlines, Air France, UK airlines and Irish Airlines, Qatar Airlines, Ethiopian Airlines, Canadian Airlines, Air India, Southwest Airlines, etc., yet the rest of the studied documents titles vary considerably, including: "Environmental Policy Statement" (South African Airways, Icelandair, Thomas Cook Airlines), "Environment Strategy" (Australian Airlines, Lufthansa, Alaska Airlines), "Environmental Mission Statement" (most of the US local airlines, like Maryland Airlines, for example), "Environmental Focus" (Scandinavian Airlines), "Environment Responsibility Statement" (Arabian Airlines), "Environmental Guidelines and Action Plan" (Japan Airlines) and "Program for sustainability" (Qantas). Some documents were formulated to cover all safety issue of the company, for example, the name the program document for Turkish Airlines is "Occupational Health and Safety and Environment Policy" and "Safety, Health and Environmental Policy" for Malaysia Airlines.

The variety of names is a rather a disadvantage as it shows that there is no clear understanding of the value of the environmental problems and the essence of those issues, which should be covered by it. It is also a losing option to combine occupational and environmental safety, as is typical for Asian and post-soviet countries.

However, the analysis of their content demonstrated it to be similar across companies, proven by the following diverse statements:

1. *Compliance with the environmental regulations and standards.* Most airlines start their environmental policies with the commitments to comply with corresponding national and international laws in the field of nature protection; however, this is a compulsory requirement for industrial activity and services provided in any civilized country. Therefore, it cannot be considered a valuable element of the environmental policy, and it decreases the importance of any responsibilities presented further in the document.
2. *Climate impacts mitigation.* Most airline companies have invested efforts to reduce greenhouse gases emissions, as it was considered a major environmental issue of modern aviation.
3. *Reduction of noise pollution.* As it was discussed before, noise is perceived almost as the most important problem of aviation activity. Even though it is hard to manage this impact at the source (aircraft on flight), the aim is still valid for environmental projects development.
4. *Protection of local communities.* Most airlines are ready to work on improvement of on-land operations to reduce the emissions of major air pollutants and noise generation by optimization of

routes, substitution of older equipment with low- and zero-emission units and restriction of auxiliary operations.

5. *Waste reduction.* Waste always imposes threats to the environmental safety and sanitary standards of any community; moreover, they are obvious problems to any active and potential customer. Therefore, improvement of the waste recycling efficiency is also among the most typical obligations taken by the airlines; however, it is related only to the waste generated during the flight. A variation of this provision is efficient management and utilization of wastewaters.
6. *Energy and resources conservation.* Energy and resource efficiency practices dedicated to save materials and consumables provide financial benefits to any company, and therefore, these aims are a top priority of most environmental strategies. It is often the only position provided with the detailed target values and plans of implementation.
7. *Information communication.* The end of most processed documents was the obligation to make employees, customers and partners more aware of the environmental issues mentioned above and encourage them to do the same.

There was a range of very specific provisions in certain policies, and this is typical for airlines of active international transportation, but not for the leaders of the market. Thus, for example, Turkish Airlines as from their policy information consider sustainable tourisms and environmental performance equally important, and they also covered occupational safety, a practice that should be covered in the separate document, not in the environmental policies section. Additionally, Turkish Airlines plan a gradual transition to alternative fuels to reduce the environmental degradation and consumption of non-renewable resources.

Some airlines have chosen to formulate their values and principles as a guarantee of human rights, combat any form of exploitation and respect to host communities (Canadian Airlines).

Scandinavian Airlines, SAS, deliver a specific element to environmental strategy, the so-called green approach. This approach intends to manage on-land operations, so that the impact on the ground layer of the atmosphere is minimal. Thus, the pilot refrains from performing any preflight manipulations until he gets the flight path and landing clearance. Using the shortest possible flight path, minimal holding in the air, optimal descending point from the cruising altitude to the runway and a slightly lowered speed, they are able to reduce fuel consumption and as a result lower emissions to the atmosphere.

Singapore Airlines have chosen to include a compensation program in their environmental policy. They have taken the responsibility to invest part of their profit to support on a regular basis the Harapan Rainforest Initiative, a collaboration between international non-governmental organizations including BirdLife International and Burung Indonesia, to conserve and protect the rainforest with one of the highest biodiversity in the world. The conservation of biodiversity is also included in the environmental policies of Japan Airlines, Alaska Airlines and Australian Airlines.

Australian Airlines are highly interested in rational usage of scarce water resources, which is a typical problem for the continent, so they formulated the requirement to reduce water consumption and install water-efficient appliances at all partner organizations.

The distinguishing trait of US airlines environmental policies is the low priority on climate-related impacts of aviation, which is considered the most significant negative impact of aviation by European companies. As a result, the US companies focus on local air quality and noise. Because of uncertainties regarding the real radiative forcing of greenhouse emissions and the impacts of aviation particulate matter on the global dimming, force air transport companies undervalue these issues. Yet they are still included in the action plans and environmental commitments (American Airlines, Delta, etc.).

Another important issue for major US airlines is increased attention to the land-use efficiency. The urgency of this issue has risen since federal authorities have postponed or stopped plans of development of the top American airports due to concerns about the local air quality and noise pollution. The similar target is set by Alaska Airlines, which strive to conserve the natural environments in wild condition with the slogan: "We fly people to beautiful places; let's keep those places beautiful."

Some companies, especially small ones, include only one clear target in their environmental policies, and this is climate effects reduction (Qantas [Australia], Thomas Cook Airlines [UK], etc.). We consider this a good approach as it gives possibility to achieve noticeable success with limited resources and directed to one specific goal. It would be even more efficient if the policy included a strict period and indicators of progress, so that in the right time, new targets could be introduced.

Southwest Airlines included in their policy aim to prevent reagents and fuel spills, which is not present in companies' policies. This impact was mentioned among those that are of importance for the environmental performance of the airlines, and lack of attention to this threat is a sign of lacking awareness about the environmental consequences of minor losses of fuels. It is highly recommended to pay attention to this issue, especially accounting for the fact that modern technologies enable maximal mitigation.

South African Airways have introduced the initiative to reduce environmental effects (especially biodiversity reduction) of food produces that supply the airlines. Since food production is among the top budget-forming sectors for airports and airlines, this commitment should be of specific interest to all operators.

The most significant targets are predictably set by major airlines, like Lufthansa or British Airways, which can plan and implement high-cost technological innovations on the way to sustainable transportation. This includes modernization of the fleet and investments into aircraft improvement, which could possibly cut carbon emissions and even decrease emissions of nitrogen oxides to some extent. Management of the company has also announced through the environmental policy the plan to optimize the route network, apply new piloting techniques that reduce noise and emissions and introduce resource-saving processes and technologies and substitution of outdated energy-intensive aircraft types on fuel-efficient ones.

From our point of view, important elements of Lufthansa's environmental policy are the introduction of monitoring and analysis of operational activities and technological processes in order to identify new opportunities for improving the environmental performance. They also plan to motivate employees to the careful spending of all types of resources.

The developed and implemented energy-saving measures are planned to save up to 1.5 million tons of aviation fuel per year and reduce its specific consumption by 43.6% by 2020.

The second-largest airline in Germany Air Berlin (Airberlin) was among the pioneer in the field of environmental protection in EU with their program "The Ecological Flight," for which Airberlin in 2011 was awarded the ÖkoGlobe award for environmental mobility. Having achieved fuel consumption of 3.5 liters per 100 passenger-kilometers, Airberlin became the first among European airlines.

Airberlin constantly invested in the latest technologies, and the average age of their aircraft was 5 years old; it is one of the youngest, least noisy and highly efficient fleets in Europe. By contrast, the average age of airplanes in the world is about 14 years and in Europe—12 years (IATA).

The program for ensuring efficient fuel consumption included 44 measures. The company believed that fuel economy started long before the engines switching on and developed special training program for their pilots. In addition, Airberlin was the first German airline to perform precision satellite-based landing with the RNP-AR method (Required Navigation Performance—Authorization Required) in Innsbruck and Salzburg. This way pilots made the flight trajectory optional and fuel saving. Unfortunately, Airberlin went bankrupt and its progressive solutions are currently not employed.

The Airbus airline and Canadian airline Air Canada, together with the BioFuelNet Canada Laboratory at McGill University (Montreal, Canada), have defined their main environmental target as the development of alternative fuels with lowered level of CO_2 emissions. This is a long-term plan and investment; however, it must be supported with a range of short-term projects, which can bring environmental benefits on a regular basis and a closer timescale.

Some companies develop very ambitious environmental programs aimed at complete transformation of the current industry practices, including significant reduction of greenhouse emissions ("Eco-skies" by United and "Sky Eco" by Japan Airlines). They include both technological and organizational solutions.

The environmental policies adopted by the major airlines in the countries of the NIS are similar by their content and mostly represent the analogues of the policies developed by major European former airlines. Thus, the environmental policy of Aeroflot, an undisputed leader of civil aviation in Russian Federation and the whole region, is aimed at improving the energy efficiency, materials conservation and the implementation of the European CO_2 monitoring and accounting system in accordance with the methodology adopted under the EU ETS (European Emission Trading Scheme for Greenhouse Gas Emissions).

Air Astana stated in their environmental policy that noise reduction activities are of primary importance. For this purpose, they developed and implement the special program of routing optimization, which will provide fuel and air pollution consumption reduction.

6 Discussions

The development and implementation of an environmental policy are an important step toward the sustainability of aviation operations in the long-term and short-term perspective. But the real success will come if the corresponding documents are perceived as valid action plans with clear target points and efficient instruments. However, the analysis has shown a range of serious obstacles for the considered environmental policies to become road map to environmental safety of airlines.

First of all, we must highlight that most policies are compiled from general provisions without any specific figures or benchmarks. The exceptions are Lufthansa group, South African Airways, Japan Airlines, Scandinavian Airlines and Irish Airlines. Such formulations create the situation when companies do not have to conduct the analysis of their achievements and failures and do not amend or adapt their policies to changing environment. This is fatal for the efficient environmental management; continuous control and improvement are the basis of efficient EMS and the only way to provide the best results of environmental impacts reduction. For example, one of the reports by Scandinavian Airlines state that the result of their innovation saves around 150 kg per landing of fuel with a Boeing 737-600, equivalent to the reduction in CO_2 emissions of over 450 kg. Lower flight speed on flights between Oslo and Bergen since 2008 showed that carbon dioxide emissions have been reduced by 420 kg (Scandinavian Airlines 2013).

Another important fault of the policies is that the work on minimization of greenhouse emissions is often reduced to the offer to passenger carbon offsetting certificates. The implementation of voluntary carbon offsetting schemes relocates the responsibility for climate stabilization from generators of impact, which are airlines, to free will of passengers. This scheme will not give visible result as numerous studies show that passengers, even highly educated passengers who generally have environmentally friendly attitude, are disinterested in increasing the cost of their trips by compensating the impacts on the environment (Hares et al. 2010; Lassen 2010). Moreover, the emission calculation methods and offsetting prices are calculated with various methodologies, which have no single standard. This leads to a reduced credibility of this approach on the whole (Hofer et al. 2010).

The trend to compensate the impacts, instead of improving operations, is seen in various environmental policies, not directly connected with transport operations. These are initiatives on financial support of local protected areas and vulnerable communities. Of course, the value of the intentions of this kind is undisputable, but the results will not lead to more sustainable airlines.

A rougher approach is to optimize free luggage allowances and to involve airlines into carbon offsetting projects. The first is uncomfortable for passengers, while the second is a problem for airlines, as they will have to bear most of the financial burden for emissions. However, in the later case airlines will have a good stimulus to find real answers to the question on how to reduce financial losses brought by impacts on the environment.

Real environmental friendliness is the last option as it stimulates the airlines to create practices and lead to improved technologies and aircraft structures.

Concentration on climate and noise issues leads to omitting serious threats to the environment, such as soil and waters pollution, biodiversity loss and habitat degradation. The result is a devaluation of environmental policy; by striving for global but hardly achievable goals, companies lose the possibility to make meaningful changes on the local level—where it really matters—and thus create a situation where they have nothing to accomplish. It is recommended to set clear targets with specified deadlines (normally 3–5 years) and measurable criteria of progress and efficiency. To make the policy work, one should plan and implement regular monitoring of performance and achievements to continuously improve the policy.

There is still one serious problem with environmental policies: Even though they present the advantageous image of an airline company, researches show that it has low influence on purchasing practice of customers (Araghi et al. 2012). Thus, the studies show that people are not ready to pay more for the same trip or to choose certain airlines only for their better environmental performance and presence of environmental policy (Hares et al. 2010; Lassen 2010). There are some minor differences among the respondents. Wealthier people, females and elder people are more interested in environmental issues and may consider this fact when choosing a transportation company, but on the whole, most customers do not consider airlines environmental performance when buying tickets, just as they are not willing to pay carbon offsetting fees (Araghi et al. 2012; Kadhi 2008). We believe this is conditioned by a low awareness about direct impacts of aircrafts on the environment, which could potentially affect people all over the world and not only local communities, as it is in the case of global climate change against local noise pollution. Of course, such ignorance is also the result of the passengers' personal attitude toward the environment as they can simply neglect it.

Still, the efforts aimed at the improvement of the level of environmental culture and education will finally form environmental consciousness of high quality, and eventually, these issues will become an important element of companies' image to attract passengers, partners and investments. It will also help to decrease negative attitude of local communities to the projects of airports expansion.

7 Conclusions

The aviation industry generates a variety of environmental impacts, including global and local climate changes; pollution of the ground layer of the atmosphere, soils and surface waters pollution on the local level; noise pollution of the airport and adjoining territory; habitats fragmentation and loss of diversity; reduction of fertile soils and lands; consumption of water and energy resources; electro-magnetic pollution; and health effects, among others.

To combat these negative environmental issues, an environmental management strategy must be formulated and presented to the public in the form of an environmental policy. Many airlines have already developed the corresponding document under variety of titles. The targets set in these documents cover very similar environmental issues, including climate effects, local air and noise pollution and resources conservation. But we have defined that the first goals are often unclear and populist, while the last one is the airline's highest priority as it gives financial benefits, yet low local benefits.

Many companies have included specific provisions in their policies, like support of protected areas or support of sustainable tourism, but the analysis shows that efficiency of these policies is not high enough and is not correlated with the improvement of environmental safety of aviation.

Another important disadvantage is the absence of progress control procedures, lack of a numerical presentation of targets set and lack of indicators for the evaluation of their achievement. It is also recommended that they pay attention to the nature protection strategies of issues such as underground waters and soil protection from the pollution caused by fuel spills and leakages, as well as unregulated application of reagents. Companies must quit declaring impossible targets or implementing compensation, instead of changing their technologies and practices.

The problem of motivation for companies to implement open public plans of achieving sustainability is partially conditioned by the population's low interest in environmental policies. Technology has succeeded in reducing aircraft emissions by 40% in engine improvements and 20% in reduced air resistance of the fuselage (Kharina and Rutherford 2015). However, the growth of airline industry raises new environmental demands to the industry and new goals, so new policies must arrive.

References

Bennis, W., Parikh, J., & Lessem, R. (1995). *Beyond leadership: Balancing economics, ethics and ecology (developmental management)*. Cambridge: Blackwell.

Brandrup, J. (2005). Polymers, polymer recycling, and sustainability. In Anthony L. Andrady (Ed.), *Plastics and the environment* (762pp.). Toronto, ON: Wiley. https://doi.org/10.1002/0471721557.ch13.

Čančer, V. (2000). Environmental management of business processes. *Management, 5*(2), 79–93.

Cherniak, L., & Radomska, M. (2017). An analysis of local climate effects: A case study of the Lviv International Airport. *Studia Periegetica, 1*(17), 123–138.

Clarke, R. A., Stavins, R. N., Greeno, J. L., Bavaria, J.L., Cairncross, F., Esty, D. C., et al. (1994). The challenge of going green: A perspective on environmental regulation and competitiveness. *Harvard Business Review, 70*, 38–39.

Collados, C., & Duane, T. P. (1999). Natural capital and quality of life: A model for evaluating the sustainability of alternative regional development paths. *Ecological Economics, 30*(3), 441–460.

DeSimone, L. D., & Popoff, F. (1997). *Eco-efficiency: The business link to sustainable development*. World Business Council for Sustainable Development. London, UK: MIT Press.

El Kadhi, W. (2008). *Do southern English airline passengers consider airlines' environmental policy in their purchase decision making?* (109pp.). Hamburg: Druk Diplomica Verlag GmbH.

Hares, A., Dickinson, J., & Wilkes, K. (2010). Climate change and the air travel decisions of UK tourists. *Journal of Transport Geography, 18*(3), 466–473.

Hofer, C., Dresner, M., & Windle, R. (2010). The environmental effects of airline carbon emissions taxation in the use. *Transportation Research Part D: Transport and Environment, 15*(1), 37–45.

Ir. Yashar Araghi, Dr. ir. Maarten Kroesen, Dr. ir. Eric Molin, & Dr. ir. Bert vanWee. (2012). *Examining consumer preferences on environmental policies designed for sustainable airline industry* (14pp.). Delft: TRAIL Research School.

Kharina, A., & Rutherford, D. (2015). *Fuel efficiency trends for new commercial jet aircraft: 1960 to 2014* (27pp.). Washington, DC: International Council on Clean Transportation.

Lassen, C. (2010). Environmentalist in business class: An analysis of air travel and environmental attitude. *Transport Reviews, 30*(6), 733–751.

O'Callaghan, P. W. (1996). *Integrated environmental management handbook* (386pp.). Chichester, UK: Wiley.

Reijnders, L. (1996). *Environmentally improved production processes and products: An introduction*. Dordrecht: Kluwer Academic Publishers.

Ruffing, K. G. (1998). Sustainable consumption and production. *UNEP Industry and Environment Review, 21*(4), 12–15.

Scandinavian Airlines. (2013). *At the forefront of the airline industry. Report on the 5 years progress* (12pp.). Scandinavian Airlines: SAS.

Tie, X., Brassuer, G. P., Granier, C., De Rudder, A., & Larsen, N. (1996). Model study of polar stratospheric clouds and their effect on stratospheric ozone: 2. Model results. *Journal of Geophysical Research, 101*(D7), 12575–12584. https://doi.org/10.1029/96JD00403.

Travis, D. J., Carleton, A. M., & Lauritsen, R. G. (2002). Contrails reduce daily temperature range. A brief interval when the skies were clear of jets unmasked an effect on climate. *Nature, 418*(6898), 601.

Ulhoi, J. P., Madsen, H., & Rikhardsson, P. M. (1996). *Training in environmental management—Industry and sustainability, part 1: Corporate environmental and resource management and educational requirements.* European Foundation for the Improvement of Living and Working Conditions, Loughlinstown, Dublin.

Walley, N., & Whitehead, B. (1994). It's not easy being green. *Harvard Business Review, 72*, 46–52.

Welford, R. (1994). *Cases in environmental management and business strategy.* London: Pitman Publishing.

Winter, G. (1995). *Blueprint for green management: Creating your company's own environmental action plan*. London: McGraw-Hill.

CHAPTER 6

Environmental Policies in European Aviation: A Stakeholder Management Perspective

Marina Efthymiou and Andreas Papatheodorou

1 Introduction

Air transport improves accessibility and social inclusion, thus enhancing direct investment and generating positive spill-over impacts on the production capacity of an economy (Papatheodorou and Lei 2006; Liasidou 2017). Interestingly, the number of flights in Europe has risen by 80% between 1990 and 2014 and is forecasted to expand by a further 45% between 2014 and 2035 (EASA et al. 2016); in this context, Boeing (2017) also forecasts the need for a total of about 40,000 aircraft over

M. Efthymiou (✉)
Business School, Dublin City University, Dublin, Ireland
e-mail: marina.efthymiou@dcu.ie

A. Papatheodorou
Department of Business Administration,
University of the Aegean, Chios, Greece
e-mail: a.papatheodorou@aegean.gr

Department of Business and Law Studies,
University of Bari Aldo Moro, Bari, Italy

T. Walker et al. (eds.), *Sustainable Aviation*,
https://doi.org/10.1007/978-3-030-28661-3_6

the next 20 years, 23,470 of which will provide additional seat capacity in the market.

At the same time, the rapid air transport growth has created serious environmental concerns ranging from noise pollution to climate change; this may substantially undermine aviation development in the future. The estimated regional share of Europe's carbon dioxide (CO_2) aviation-related emissions without a regulatory intervention is expected to reach 36.60% in 2020 (ICAO Doc 10018, 2013). Moreover, the European Commission (2018) states that by 2020, global international aviation emissions are projected to be around 70% higher than in 2005, and by 2050, they could grow by a further 300–700%.

Many climate change mitigation policies (e.g. the 2015 Paris Agreement) are currently implemented throughout the world and become an integrated part of the wider policymaking agenda. Nonetheless, aviation is not regulated in comparison with other industries in terms of its negative environmental externalities. The gas emission problem is aggravated by air traffic delays and congestion that extend the flying time and route, which in many cases relate not only to limited airport infrastructure, but also to airspace fragmentation as experienced at least in Europe. Efthymiou et al. (2016) highlight the environmental implications of such delays for the travel business ecosystem, which may build on airport-related delays to affect negatively the overall passenger experience and satisfaction (Efthymiou et al. 2019). Surprisingly, however, there has been little academic discourse on the management aspects of aviation environmental regulation, especially regarding Air Navigation Service Providers (ANSPs), i.e. public and/or (semi-)private entities in charge of air traffic management (ATM) usually at a national level.

In his state-of-the-art discussion of environmental ethics in tourism, Holden (2018) stresses the importance of active stakeholder engagement in shaping a new Weltanschauung and morality vis-à-vis mobility; at the same time, he acknowledges that dealing with air travel and the right to fly is 'ethically complex' as environmental taxes on aviation (just like hotel eco-taxes) may end up hurting the lower socio-economic classes to the detriment of democratisation of travelling. Therefore, issues of power and interests in the context of multi-stakeholder governance become of essence especially when the various policy schemes, aiming at alleviating the problem, are characterised by significant transaction costs in their implementation and verification.

On these grounds, this paper aims to contribute to the aviation literature by investigating and addressing governance failure in the target-setting and the implementation process of both demand- and supply-oriented environmental schemes in aviation. More specifically, the EU Emissions Trading Scheme (EU ETS) in aviation and the Single European Sky (SES) initiative are examined. The former aims at regulating civil aviation operations by internalising negative external economies related to climate change: essentially, the scheme manages demand by creating a new market for tradable emission permits. As a result, the EU ETS poses several challenges (but also potential opportunities) for airlines, which now see their costs rising. The latter, i.e. the SES reform, focuses on restructuring the airspace system and reshaping the Air Navigation Service Providers (ANSP) sector. SES deals with congestion problems and traffic inefficiencies in aviation from a supply-side perspective. In fact, reduced inefficiencies lead to shorter flight times and lower levels of fuel consumption and CO_2 production thus having positive environmental effects. Unlike the EU ETS, its introduction has been welcomed by the airlines, but as expected, it raises new challenges and possibly resistance to change from other stakeholders especially within the ANSP sector.

To the best of the authors' knowledge, this is the first research attempt to consider SES and EU ETS simultaneously. A joint policy approach is followed to investigate the standpoints of the different stakeholders and consider managerial implications regarding the environmental regulation of aviation. This is because a potential misalignment between the EU ETS and SES may prove problematic. For example and as discussed later in the chapter, airlines may decide to fly longer routes to avoid airspace segments characterised by high air navigation charges thus increasing fuel consumption and CO_2 emissions: this could make financial sense if the cost of acquiring CO_2 permits is low but undoubtedly has negative repercussions on the environment thus cancelling the very introduction of the EU ETS scheme. Therefore, an effective communication between those managing and regulating the two schemes is necessary to streamline operations and avoid negative side effects.

Section 2 of the chapter provides background information on EU ETS and SES. From a theoretical perspective, Sect. 3 discusses the appropriateness of stakeholder analysis supported empirically by the focus group study, the observation of participants and the conduct of interviews. Section 4 then reports and discusses the empirical results and their policy and stakeholder management connotations while Sect. 5 summarises and concludes.

2 EU ETS in Aviation and SES: An Overview

2.1 EU ETS in Aviation

The EU ETS is based on the requirements set by the Kyoto Protocol (KP). The EU ETS is a 'cap and trade' scheme. The carbon trading involves trading of rights (permits, allowances, credits) to emit a certain amount of emissions. Because of this commercial dimension, an economic incentive is created (Preston et al. 2012). A quantity of allowances (also called permits) to emit one metric tonne (MT) of CO_2 is given for free and excess permits need to be bought by each market participant. At the end of each year, each entity must return allowances at least equal to its level of emissions for that particular year. In case the entity does not return enough allowances, it incurs a fine. As time passes by, the emissions ceiling will be reduced and fewer allowances will be issued, causing CO_2 emissions to be reduced too (Leggett et al. 2012; Meleo et al. 2016). In 2008, the European Parliament and the Council adopted Directive 2008/101/EC, amending the EU ETS (Directive 2003/87/EC) to include aviation activities. The EU ETS in aviation includes only CO_2 emissions from aircraft (Kantareva et al. 2015). Every MT of CO_2 emitted by a source should be covered by a corresponding right/allowance. Allowances are initially allocated through a benchmarking system regarding the tonne-kilometre (tKm) distance flown by an aircraft from, to or within the EU. Every participant receives 0.6797 free allowances (baseline) for every 1000 tKm of flight. The smaller the quantity of free allocated allowances is, the bigger the associated cost for aircraft operators (IATA 2013). Moreover, to minimise compliance costs and to provide flexibility to airlines, the EU ETS permits allowances to be traded (Schleich et al. 2006; Tietenberg 2010). Airlines requiring more allowances can purchase them from EU auctions, other carriers and other entities within EU ETS or other international emissions' trading mechanisms. In 2012, 85% of emission allowances in aviation were offered free-of-charge to aircraft users and 15% were auctioned (European Commission 2016). For the period 2013–2020, the free allocation is further reduced to 82% and 3% of the allowances are banked for new entrants in the market and/or rapidly expanding airlines (European Commission 2016; Anger and Kohler 2010). The EU ETS excludes certain flight types from the cap-and-trade system, such as flights from airports located outside the EU, military aircraft, and Public Service

Obligation (PSOs) routes (Directive 2008/101/EC). Furthermore, flights under the de minimis criteria are excluded (Kantareva et al. 2015). In February 2018, EC Regulation 2018/208 amended Regulation 389/2013 to prepare for the case of a hard Brexit and ease the possible withdrawal or auctioning of allowances.

From a theoretical perspective, the application of EU ETS in aviation has both neo-liberal and interventionist facets. The former is based on the Coase Theorem in welfare/environmental economics, while the latter is typically related to Pigouvian taxation aiming at correcting negative production externalities (Stabler et al. 2010). The Coase Theorem essentially argues that when property rights are well defined, markets are competitive, and transaction costs do not exist, then the level of the externality (i.e. emissions) is set irrespective of the actual allocation of property rights (i.e. tradeable emissions permits) as agents with quasi-linear preferences reach a mutually beneficial outcome (Coase 1960, 1992). In the case of the EU ETS, the total level of emissions is capped externally (i.e. by the European Union) and not determined by market forces. In other words, a new market for emissions is created and the EU allocates tradeable permits and intervenes, as in the case of other environmental taxes, to ensure that airlines internalise their externality. The entire EU ETS mechanism is rather decentralised implicitly assuming that the very assumptions of the Coase Theorem are met to ensure an optimal outcome. Nonetheless, property rights in aviation may not be as well defined as originally thought, and transaction costs in carbon offset schemes may occur for various reasons. In particular, the rising concentration of the European airline market may not only raise concerns about the competitive conduct (Németh and Niemeier 2012), but also lead to protracted bargaining between the leading airlines and the EU ETS competent authorities. Moreover, local communities may resist against new airport developments (Santos et al. 2017) and also challenge the regulator's competence to oversee the scheme. All the above may possibly challenge the relevance of the Coase Theorem in the EU ETS leading to the conclusion that '...compensation through initial allocation would no longer be emissions neutral' (Abrell et al. 2011, p. 2).

From an interventionist standpoint, fuel taxation could be a simpler measure to monitor compliance with lower transaction costs, as fuel consumption is easy to measure and therefore tax, whereas EU ETS is a more complex system with free allocation, carbon offsetting principles and permits trading. Nevertheless, international aviation fuel taxes are not permitted under various bilateral agreements and under Council Directive 2003/96/EC (Korteland and Faber 2013). The latter allows EU member states to tax aviation fuel for domestic flights and by means of bilateral agreements. Taxation levels should be below the minimum set out in the Directive. According to Korteland and Faber (2013), an imposition of aviation fuel tax of €330–350 per MT by the EU would generate revenues between €20 and 32 billion. If a VAT of 20% on jet fuel were introduced, there would be revenues between €10 (based on current fuel prices) and €14–16.5 billion (prices plus fuel tax). In any case, if VAT were imposed, airlines could deduct it against the VAT receipts from airline tickets (Korteland and Faber 2013). As with the case of the Balearics ecotourism tax introduced in 2016 (and modified later in 2017), airports and local communities have adopted, or tried to adopt, measures to regulate airline operations regarding local air quality by implementing Pigouvian taxes. Such taxes (like the Catalonian NO_x tax) are not well received by airlines which often threaten airport authorities and/or local governments of moving their operations to other airports in the vicinity governed by a more relaxed tax regime. Pigouvian taxes aim at bringing marginal private costs in line with social costs that cannot be accurately measured. Accurate measurability, which is usually missing, leads to improper corrective steps. Therefore, taxes are not a very successful example of environmental regulation in the European aviation market.

The implementation of the EU ETS policy in aviation has raised tension among the government and industry stakeholders at an international level. Representatives of the US government and of other countries (such as fast-developing countries like China) claimed that the horizontal implementation of the EU ETS to non-European carriers violates the 1944 Chicago Convention of the International Civil Aviation and its bilateral agreements and that the issue should be resolved by ICAO (Havel and Sanchez 2012). Some US airlines and Airlines for America (A4A) appealed to the European Court of Justice (ECJ). However, the latter concluded that the EU ETS did not infringe the Chicago Convention, the KP, the US-EU Open Skies Agreement, the principle of territoriality or the sovereignty of third-party states (Court of Justice of the European Union 2011). There are countries that are positive towards the full implementation of the EU ETS scheme because they have a commercial interest in biofuels. Brazil is one of them, since Boeing and Embraer opened in 2015 a Joint Aviation Biofuel Research

Centre in Brazil (Embraer 2015). This example illustrates that among stakeholders, there are subgroups that are either supportive or against the scheme according to their interests. As a result of this tension, the EC decided to 'stop-the-clock' (Decision No. 377/2013/EU) regarding the EU ETS in 2013. According to this Decision, air carriers that depart or land to a non-European Economic Area (EEA) airport are not obliged to surrender any allowances back and are exempted from the EU ETS. Moreover, in 2016, ICAO established the Carbon Offsetting and Reduction Scheme for International Aviation (CORSIA), aiming to stabilise CO_2 emissions at 2020 levels. According to ICAO Assembly Resolution A39-3 (2016), CORSIA will become operational in 2021; the pilot (2021–2023) and the first phase (2024–2026) will be voluntary (all EU countries will join in 2021), while participation in the second phase (2027–2035) will become mandatory for all ICAO states. CORSIA is an offset scheme with no cap that does not aim to achieve reduction of emissions below their 2020 level (Scheelhaase et al. 2018) while the EU ETS aims to reduce further its cap. The geographical scope as well as the ambition for reducing CO_2 emissions is substantial differences of CORSIA and EU ETS. The EC decided to retain the current intra-EEA system at least until 2021.

2.2 *Single European Sky*

The European ANS system covers 37 ANSPs, 60 control centres and a territory of 10.8 million km^2, and according to the European Commission (2017), the estimated costs of fragmentation of airspace amount to 4 billion EUR a year. The European sky is fragmented in small inefficient blocks that use a wide variety of Air Traffic Control (ATC) technology (European Commission 2017). For instance, a flight from Athens to Rome is serviced by the Greek ANSP and the Italian ANSP Traffic that follows different rules and operational requirements and use different equipment/systems and support functions. This fragmentation results in mis-coordination and mis-management of the airspace and therefore in safety risks, delays in flights and increased cost. Thus, better coordination for transferring the responsibility of an aircraft between two ATC sectors in Europe is needed. Increased traffic, capacity problems, climate change and the necessity of developing a more cost-efficient system led to an ambitious initiative to reform the ATM. This initiative was first launched by the European Commission in 1999 and is called Single European Sky (SES).

One of the core elements of the SES initiative is the Functional Airspace Blocks (FABs). According to Regulation (EC) No. 1070/2009, a FAB is an airspace sector based on operational requirements. It reflects the need for integrated management of the airspace regardless the existing boundaries of Flight Information Regions (FIRs), where the provision of air navigation services and related functions is oriented towards performance and is optimised to implement a cooperative relationship between the ANSPs in each FAB.

According to the European Commission (2012), SES aims to improve safety by ten times, triple the capacity of airspace and reduce the cost of ATM by 50% and the impact on the environment by 10%. The performance in SES focuses on four Key Performance Areas (KPAs), i.e. safety, capacity, cost-efficiency and environment. The implementation of the Performance Scheme aims at setting and implementing binding targets for EU member states through the adoption of EU-wide performance targets and approval of consistent national or FAB performance plans.

The main metric of the environment KPA is the average horizontal en-route flight efficiency indicator, which is the difference between the distance of the en-route part of the trajectory and the optimum trajectory, that is, on average, the Great Circle Distance (GCD) (Guastalla 2014). Thereby, 'en-route' is defined as the distance flown outside a circle of 40 Nautical Miles (NM) around the airport. Flight efficiency can be measured horizontally or vertically, and during the last decade, more efficient routes have been implemented. Currently, the European route network distances are 3.6% longer than the GCD (IATA et al. 2008).

Better route network design as well as airspace availability and utilisation can improve flight efficiency (Efthymiou and Papatheodorou 2018). Restrictions imposed on the European route network utilisation create approximately 0.4% airspace utilisation inefficiency defined as deviation from the GCD. For instance, a DANUBE FAB Environmental Impact Assessment Study was carried out in 2011 using the System for traffic Assignment and Analysis at a Macroscopic level (SAAM) fast-time simulation tool to calculate the changes in fuel use and CO_2 emissions (Kantareva et al. 2015). The study concluded that the annual fuel saving due to the FAB implementation would be 45,000 tonnes by 2020 and 80,000 tonnes by 2030. The annual CO_2 savings are expected to be 143,000 tonnes by 2020 and 255,000 by 2030 (Kantareva et al. 2015). The environment KPA regulates all aircraft emissions that have an impact on climate change in comparison with EU ETS that regulates only CO_2 emissions.

In contrast to EU ETS, airlines welcome the introduction of SES as this does not increase their production costs; nonetheless, changes in the European ATM are often resisted by the powerful ANSP labour unions who fear that the unification of the European skies may potentially lead to significant job losses in a sector characterised by high levels of specialisation (Blondiau et al. 2016). The very decision of the European Union to head for the creation of different FABs instead of a 'real single sky' is consistent with an effort to balance the conflicting interests of different stakeholders. However, such an approach inevitably raises transaction costs in the system thus also endangering the attainment of environmental objectives as also discussed in the empirical sections of this chapter.

3 Empirical Methodology

The implementation of environmental regulation poses significant challenges for the various involved parties. For this reason, a stakeholder analysis is deemed appropriate to understand and analyse the issue holistically (Farmaki and Papatheodorou 2015); after all, managing relationships to ensure that stakeholders behave according to the SES and EU ETS objectives is of critical importance for social welfare. Stakeholder analysis is conducted to assist institutional and policy reform processes by elucidating and often addressing the demands of those who have an interest in the reforms under question (Moloney 2006; Pigman 2007). Freeman et al. (2010, p. 29) argue that 'this approach is consistent with the main ways in which we understand capitalism', namely neoliberal economics (Friedman 1962), principal-agent issues (Jensen and Meckling 1976), strategic management (Porter 1985) and transaction costs theory (Williamson 1975). In fact, the essence of both EU ETS and SES relies on market regulation principles, which accept the need to secure a decentralised equilibrium based on the comparative and competitive advantage of the aviation supply chain participants (and predominantly of airlines) albeit in the presence of monitoring, reporting and verification costs as well as constraints related to information asymmetry. Moreover, to encapsulate the stakeholder mindset from an empirical point of view, a combination of methods involving focus group, participant observation and the conduct of interviews is pursued here. Such a combination steps beyond the traditional positivist approach and delves into interpretivism and social constructionism (Saunders et al. 2015), which are of essence when trying to understand the background behind the expressed (expert) opinions.

The current focus group analysis is part of a wider research project and is conducted in two-round interviews to encapsulate policy complexity at an in-depth level. The target population comprises stakeholders involved in SES and EU ETS. To select candidates, a purposive and snowball sampling was used. Thus, the sample of the wider research consists of thirty-nine (39) senior managers/experts from Civil Aviation Authorities (CAAs), ANSPs, aviation-related organisations and institutions and airlines. This research approach reinforces the literature review and leads to the identification of related transaction costs and the development of a stakeholder power vs. interest grid (Eden and Ackermann 1998) regarding environmental regulations in aviation.

Participant observation at EUROCONTROL was conducted in 2015. Seven (7) semi-structured interviews were conducted that contributed also to the development of some of the focus group research questions. The seven interviewees had different backgrounds. One senior expert worked at EUROCONTROL; another at European Commission; one expert worked in the Non-Governmental Organization (NGO) Transport and the Environment; two others were senior consultants/academics in aviation and environment; and finally, two senior experts worked for the International Air Transport Association (IATA). The interviewees requested their profile to remain anonymous without stating their specific roles within their organisations.

The stakeholders face four major attributes (Stoney and Winstanley 2001), the stakeholders' position on the reform issue, i.e. SES and EU ETS; the influence (negotiation power) level; the interest level; and the group/coalition to which they are affiliated. These four attributes were determined via interviews with experts and via the actual stakeholders directly, as well as from literature review and public announcements. Moreover, the stakeholders have power over the potential formation and/or implementation of a regulation. The power is described as the capability to influence or compel the actions of others and can be coercive, utilitarian or normative (Johnson et al. 2010; Ihlen and Berntzen 2007). The level of influence depends on the stakeholders' power to promote their position on regulatory reforms. Based on the combination of the participant observation and the semi-structured interviews, the positions of the SES and EU ETS stakeholders are stated in the following section.

4 Discussion and Results

4.1 *Key Stakeholders*

According to interviewees from EUROCONTROL and IATA, it is of critical importance to understand first who the stakeholders and their positions are and then design a roadmap to effectively manage them. The main stakeholders in the EU ETS and SES reforms are the European Commission, the European Parliament and national governments, the airspace users, other organisations and institutions and the ANSPs. In this section, their positions, as interpreted from the participant observation and the semi-structured interviews, are presented.

4.1.1 Organisations and Institutions

The European Parliament is the ultimate decision-maker of the reforms and can approve, amend or reject a legislative proposal. The European Parliament recognises that it has a duty to make a positive contribution to welfare and sustainable development as a long-term goal, through both its political and legislative role. According to the European Commission interviewee, the European Commission's position coincides with the European Parliament's position. The interviewee from EC stated that the European Commission is asked to speed up the full implementation of SES and switch from a 'bottom-up' to a 'top-down' approach if needed. The interviewees from EUROCONTROL and IATA implied that the European Parliament is influenced many times by industry bodies in terms of its policy direction.

All the interviewees agreed that the regulation and oversight of SES are quite weak and argued that the member states have a 'vested interest' in maintaining monopolies of ANSPs and restraining competition. This statement is reinforced by a Performance Review Body (PRB) report (2016, p. 8) claiming that 'Unhelpful behaviours and gaming are observed. Not only in the regulated community but also in the operational elements of the legislation'. Since most ANSPs are government-owned and the charging scheme is such that the ANSP cannot make a loss, the national governments have a strong incentive to preserve the situation as it is in terms of ownership and competition. Thus, ministries retain control of the infrastructure and manage the interaction within regulating bodies leading to a clear conflict of interest. Hence, the level of independence and credibility of the regulator is determined by the prevailing political economy that establishes the regulator, sets its goals and instruments and is always able to subsequently change the rules.

EUROCONTROL is self-defined as an independent organisation. It claims that it supports and follows the regulations promoting the efficiency and effectiveness of the aviation systems for all its forty-one (41) member states equally. The IATA has the mission 'to represent, lead, and serve the airline industry' as advocated on the organisation's website. The airlines follow the developments on SES, and they want the aviation system to become more efficient in terms of both charges and operations. Interviewees from IATA and the NGO state that due to the stop-the-clock initiative, the EU ETS has a negative impact on airlines operating on European routes. The NGO interviewee finds the reduced geographical scope of EU ETS unfair for European airlines and less environmentally effective.

4.1.2 Airspace Users

The term, 'airspace users', refers to aircraft operators and especially airlines. By regulating ANSPs, SES aims to lower navigation charges and shorten routes and make better use of the airspace altogether. Hence, airlines are in favour of the SES reform of ATM. This was also confirmed by interviewees from IATA and academia. Depending on the airline's business model, the direct cost for ANS represents between 6% and 20% of the total operating costs, excluding fuel (interviewees from IATA). In addition, there are costs for delays and flight inefficiencies (due to longer routes and higher levels of fuel consumption).

Fuel is the second or the third highest cost (depending on the fuel price), and the fuel consumption causes carbon emissions. Thus, by minimising its consumption, airlines minimise their carbon footprint. However, for the minimum consumption of fuel to be achieved, airlines need to invest in new aircraft, new technologies, trainings of pilots and other measures (interviewees from academia and consultancies). Thus, the EU ETS regulation causes concern to many airlines. Most airlines oppose the inclusion of aviation in EU ETS (NGO interviewee). Nevertheless, there are some airlines (e.g. Lufthansa) that have already been following a more environmentally friendly approach and are investing a lot of capital to the environmental improvement of their operations (interviewee from IATA). Thus, those airlines embrace the EU ETS principles and concepts. Another condition that influences the position of the airline towards EU ETS is the size and ownership of the airline (interviewees from IATA). Airlines that have a small fleet and limited flight services end up spending more time and effort in EU ETS than the benefits they get back from selling unused allowances. Moreover, airlines that are government-owned need to have the approval of the related public authority in charge to introduce radical changes in their fleet or the management of their operations. This makes the compliance with EU ETS more difficult as the newer and more fuel-efficient fleets can have positive effect on emissions reduction. Finally, it is also a matter of available expertise and prevailing mentality according to all interviewees. Small airlines lack the necessary expertise to follow fuel-efficient strategies and operate sophisticated software for fuel planning operations.

4.1.3 Air Navigation Service Providers (ANSPs)
ANSPs have two diverse positions on SES and neutral position on EU ETS according to all interviewees apart from the NGO. Those ANSPs that are state enterprises/bodies, i.e. owned by the state, follow the position of the state, i.e. preservation of the natural monopoly. Other ANSPs, that are more profit-orientated like NATS in the UK, may be in favour of the reforms to be given the opportunity to grow in scale. Those services are mostly terminal control or approach control. According to the current national regulations in the EU, there is the requirement/limitation that the air traffic controllers speak the country's language especially in approach and terminal control (interviewee from academia). Moreover, to comply with the Performance Scheme rapid and drastic changes in the operations need to be undertaken and many times the ANSPs personnel are not willing to adapt (interviewee EUROCONTROL and EC). Finally, the SES requires investments in infrastructure (e.g. in communication and navigation systems) that the ANSPs budget cannot afford (all 7 interviewees).

5 Emerging Transaction Costs in SES and EU ETS

5.1 Transaction Costs in SES

In economics, a transaction cost is a cost incurred in making an economic exchange beyond the cost and price of the product or service involved. Williamson (1979) and Stavins (1995) suggest that transaction costs make economic instruments function improperly. In environmental economics, such costs are usually associated with the management of emissions' trading and the use of incentive mechanisms in performance regulation (Williamson 1998). In the context of the SES and the EU ETS, transaction costs are expected to differ among stakeholders. For instance, an ANSP is expected to incur a significant cost related to compliance, i.e. the cost of considering and adopting new systems and infrastructure. Policymakers spend a lot of effort on developing alternative policies, evaluating options and deciding what is to be implemented. The development and implementation of the regulatory scheme also bear high transaction costs, primarily because experts from different fields need to convene, discuss and agree on specific targets. Monitoring, reporting and verification also generate transaction costs. Companies need to design internal guidelines related to compliance with the introduced regulatory framework. They also need to develop long-term strategies to enforce and abate the regulations. In a nutshell and based on the undertaken empirical research, Table 1 classifies transaction costs for policymakers/regulators emerging from the environment KPAs in SES.

The focus group participants were first asked to rate the importance of transactions costs for the effective functioning of the environment KPA in the SES Performance Scheme on a Likert scale ranging between 1 (not important) and 5 (very important). In the second round of the focus group process, the mean score was 3.83 with a standard deviation equal to 0.64; the Coefficient of Variation (i.e. standard deviation divided by the mean) was 0.17. Subsequently, the participants were asked to allocate 100 points over the different categories of transaction costs ranked in descending order of importance. Participants were also given the option to add a new category. In the second round, development and implementation of the regulatory scheme received on average about 27 percentage points, followed by monitoring; strategy; reporting and verification (13%); alternative policies; and compliance measures. A small number of participants also provided additional categories (Table 2).

5.2 *Transaction Costs in EU ETS*

Regarding the EU ETS, policymakers and regulators face similar challenges and transaction costs as in the case of SES. As for airspace users,

Table 1 Transaction costs categories for policymakers/regulators emerging from SES

Categories	*Description*
Development and implementation of the regulatory scheme	• Quantification of historic emissions • Development of emission outlooks • Decision for an application rule • Measures to overcome 'frictions' and negotiation with stakeholders • Assessment of participants • Adaptation or purchase of software • Material costs set-up of organisational structures and assignment of responsibilities • Fees for information, training
Monitoring	• Design of a monitoring concept • Implementation of an internal monitoring system • Ongoing monitoring
Strategy	• Design of the strategy for National Supervisory Authorities, ANSPs • Design of the regulation enforcement procedure • Design of the abatement strategy
Reporting and verification	• Quantification of annual emissions • Compilation of an emissions report • Verification of an emissions report • Delivery of data for ex-post control
Alternative policies	• Develop alternative solutions • Evaluate the alternative solutions • Decision for the implementing policy
Compliance measures	• Identification of compliance measures • Offering recommendations and support • Decision about imposing non-compliance penalties

Source Own elaboration

these must monitor and report their CO_2 emissions, ensuring that they hand in a sufficient number of allowances to the authorities to cover their emissions. Monitoring, reporting and verification result in transaction costs whose largest part is incurred by airspace users. To lower transaction costs, many airlines have chosen to outsource monitoring service to others, e.g. EUROCONTROL. Moreover, based on the de minimis criteria discussed earlier, several small carriers are protected from incurring transaction costs emerging from the EU ETS. In any case, such costs are also closely related to the spot price of emission permits.

Table 2 Relative importance of SES transaction costs (2nd round SES questionnaire)

	Descriptive statistics				
	Min	*Max*	*Mean*	*SD*	*CV*
Development and implementation of the regulatory scheme	10	45	27.15	9.46	0.35
Monitoring	10	45	18.07	10.02	0.55
Strategy	5	60	17.69	11.09	0.63
Reporting and verification	10	35	13.19	5.90	0.45
Alternative policies	0	20	12.15	5.55	0.46
Compliance measures	0	17	9.20	4.29	0.47
Other	0	15	1.80	4.05	2.25

If the price of the emission permits is too low while at the same time the associated transaction costs are too high, then some airlines may abstain from reducing their emissions preferring to pay for extra permits instead. Based on the undertaken empirical research, Table 3 classifies transaction costs for airlines emerging from the EU ETS.

As in the case of SES, focus group participants were first asked to rate the importance of transactions costs for the effective functioning of EU ETS. In the second round of the focus group process, the mean score was 3.81 with a standard deviation equal to 0.49; the Coefficient of Variation was 0.13. Subsequently, and in terms of point allocation, implementation of emissions management category scored about 22%, followed by reporting and verification; application; and monitoring. Abatement measures received the lowest score (Table 4).

5.3 *Stakeholder Power vs. Interest Grid in SES and EU ETS*

The effective power of the industry, i.e. the degree of power that industry stakeholders hold over other groups in relation to a reform of the aviation environment, is very strong. As discussed earlier and based on the focus group analysis and semi-structured interviews, Table 5 depicts the power and interest grid of the stakeholders in the SES and EU ETS business ecosystem.

Regarding the EU ETS, airlines have high interest and relatively high power over the reforms. This was proven by the 'stop the clock' decision.

Table 3 Transaction costs categories for airlines emerging from the inclusion of aviation in EU ETS

Categories	*Description*
Implementation of emissions management	• Information, training • Assessment of obligation to participate in the EU ETS • Set-up of organisational structures and assignment of responsibilities • Adaptation or purchase of software • Material costs
Reporting and verification	• Quantification of annual emissions • Compilation of an emissions report • Verification of an emissions report • Delivery of data for ex-post control
Application (scheme design)	• Quantification of historic emissions • Development of emission outlooks • Decision for an application rule • Compilation of an application • Compilation of a benchmark where necessary • Verification of the application • Fees for annual allocation • Fees for emissions register
Monitoring	• Design of a monitoring concept • Implementation of an internal monitoring system • Ongoing monitoring
Trade	• Transactions fees (exchange fees, broker fees, clearing) • Trade and negotiation • Market observation
Strategy	• Definition of the risk strategy • Definition of the trade strategy • Definition of the abatement strategy
Abatement measures	• Identification of abatement measures • Decision about abatement measures

Source Own elaboration

Academia has high interest on EU ETS processes and outcomes, but relatively low power. The ANSPs have no interest and limited power over the EU ETS as the scheme does not regulate them. The regulators have high levels of interest and power. The airports have low power and low interest in EU ETS reforms as the regulation refers to operations outside of Landing and Take-Off (LTO) cycle. Several institutions and associations (e.g. IATA) also have high levels of interest and power regarding EU ETS due to their effective lobbying efforts. Fuel suppliers have quite

Table 4 Relative importance of EU ETS transaction costs (2nd round of EU ETS questionnaire)

	Descriptive statistics				
	Min	*Max*	*Mean*	*SD*	*CV*
Implementation of emissions management	0	50	22.66	11.09	0.49
Reporting and verification	0	45	16.86	10.28	0.61
Application	0	30	14.78	6.24	0.42
Monitoring	0	45	14.01	11.05	0.79
Trade	0	25	10.01	7.54	0.75
Strategy	0	30	9.40	6.23	0.66
Abatement measures	0	15	6.97	5.16	0.74
Other	0	10	1.40	3.07	2.19

Table 5 SES and EU ETS stakeholder power/interest grid

	High power and high interest		*Low power and high interest*		*High power and low interest*		*Low power and low interest*	
	SES	*ETS*	*SES*	*ETS*	*SES*	*ETS*	*SES*	*ETS*
Airlines	X	X						
Academia			X	X				
ANSPs	X							X
Airports			X					X
Regulators	X	X						
Organisations	X	X						
Fuel suppliers		X			X			
Manufacturers		X			X			
System providers	X							X

a high level of power due to their oligopolistic power, and high interest regarding the environmental targets set to airlines. Manufacturers have high power and high interest in EU ETS performance due to the possible impact that the scheme may have on the demand for aircraft and aircraft specifications by airlines. Finally, system providers have low interest and low power over EU ETS.

With respect to SES, airlines have high power and high interest as they are the ones benefited or harmed by the SES initiatives. The academic community, as in the case of EU ETS, has low power, but high interest. The ANSPs have high interest in the SES reform as they are the regulated entity and quite high power too due to strong trade unions and natural monopolistic power. Regulators (in the form of either Civil Aviation Authorities/National Supervisory Authorities or EC/EP) have the highest level of power and interest in this scheme. Airports benefit in terms of on-ground ATC improvements, but their role is more of an observer than of an active participant in SES developments. Fuel suppliers have high power but low interest in SES. Manufacturers have higher power than fuel suppliers do, but lower interest compared to airports or airlines. Finally, system providers have high power and high interest in SES as any changes affect their sales to ATC companies and their research and development costs.

As also discussed by Eden and Ackermann (1998), the players that have high power and high interest need to be managed closely because they are catalysts for the success of the schemes. The players that have high power and low interest need to be kept satisfied by the policymakers. The stakeholders with low power and high interest need to be kept satisfied but also informed by the policymakers. Finally, for those stakeholders with low power and low interest, minimal effort needs to be provided; however, action must be taken by policymakers when necessary. The stakeholders can lead to amendments or changes in the reforms under consideration due to their interests and their power. The most important example with major effects on the aviation reform relates to the reactions of many airlines against the scope of the EU ETS.

5.4 Policy Interaction and Stakeholder Management

Several environmental issues in aviation have been highlighted in the empirical analysis. Regarding the operational aspects, the problems of the air transport industry are the high cost of ATM service provision, the delays and related costs, safety issues and the contribution to climate change caused by excess fuel burn and emissions. Those problems are caused by the fragmentation of the ATM sector, labour and social issues, economic difficulties faced by the States, outdated technology and lack of airspace capacity. The reforms currently implemented are the EU ETS for aircraft operator and the SES for the ANSPs.

Moreover, when the ICAO implements the Global MBM Carbon Offsetting and Reduction Scheme for International Aviation (CORSIA) for all airlines, the scope and MRV of EU ETS will need to be revised. As also suggested by Scheelhaase et al. (2018), the European Commission may need to keep EU ETS for domestic flights. These are subject to the emissions reduction commitments and are included in the nationally determined contributions of the Paris Agreement that does not explicitly refer to aviation (ICAO 2017). Policymakers should approach this option taking into consideration all the industries and addressing climate change holistically. The lack of collaboration and coordination in R&D at regional and multi-national levels as well as within and across industries should be understood and rectified accordingly.

The environmental aspects of aviation should be considered in parallel with the financial situation of the airlines, ANSPs and states, and societal needs. Moreover, research and development is very costly, and, when developing environmental policies in aviation (like the use of biofuels), the needs and policies for the other means of transportation should be jointly considered. Links should be formed within Europe but also with the rest of the world.

Finally, managing the stakeholders is one of the most critical points to achieve an effective reform. The interests of the different airlines, airports, ANSPs, States, CAAs, fuel suppliers, manufacturers and policy consultants should be taken into consideration when designing environmental policies. The next step is to develop more platforms, in addition to the National Supervisory Authorities (NSAs) Coordination Platform (NCP), and move beyond the consultation process. The NSAs have reported that they find the NCP very useful as it is the only established official platform where they can congregate and discuss related matters in person, promoting the exchange of information and lesson learning. Similar platforms can be developed to encourage each group to meet separately or with other groups and contribute their ideas or express their concerns to the regulators.

The processes should be adapted to the needs and capabilities of the stakeholders in all states. Apart from this recognition of efforts, the communication of benefits and burdens as well as the fair distribution of benefits should be a priority for policymakers. Improved communication, monitoring and reporting, incentive systems and third-party review are key elements for the sustainable development of aviation through SES and EU ETS. In this way, potential conflicts among airlines, airports, ANSPs, CAAs, suppliers, policy advisers and makers will be identified and addressed. The joint consideration of different schemes, rather than their individual function, can bring operational cost savings. However, the linking should be done only if there are consistency and compatibility among the schemes in question.

6 Summary and Conclusions

This chapter presented two policy schemes that relate to the environmental performance of aviation in Europe, i.e. the EU Emissions Trading Scheme and the Single European Sky. These were evaluated in relation to the different stakeholders' point of view regarding the need and efficiency of the schemes. Although the primary data was collected in 2015, the results of the study are still regarded as topical given the European Commission's decision to retain the current intra-EEA system at least until 2021. In other words, even if new primary data was collected today (with the same panel of experts), results would most probably differ only marginally. On these grounds, the chapter provides a valid roadmap for policymakers to consider the different stakeholders' positions and influence the effective design and implementation of regulatory policies.

The implementation of SES and the inclusion of aviation to EU ETS have proven very challenging for regulators. The stakeholders of SES and EU ETS play a critical role in this context. Four major attributes determine the capability of a stakeholder to block, amend or promote regulations either alone or in collaboration with other stakeholders. Those are: (1) the stakeholders' position on the reform issue, i.e. SES and EU ETS; (2) the level of influence (i.e. negotiation power) they hold; (3) the level of interest they have in SES and/or EU ETS; and (4) the group/coalition to which they are affiliated. Hence, stakeholder analysis offers a detailed understanding of the aviation business environment taking into consideration political, economic and social elements that affect the groups' positions, the hierarchy of authority and the power among different groups.

There are big differences among the stakeholders, but also within the stakeholders. For instance, the European ANSPs operate in very diverse environments in terms of traffic complexity and variability as well as socio-economic conditions (e.g. cost of living, labour laws). The ANSPs size in Europe is also very diverse. The five largest ANSPs bear 57% of the total Pan-European ATM/Communication Navigation and Surveillance provision costs, while the five smallest represent less than 1%. Not surprisingly, therefore, the regulated parties resist to reforms. Nonetheless, if the policymakers and regulators understand better the positions and standpoints of the different stakeholders, then the implementation of the reforms will prove smoother and more effective.

References

Abrell, J., Faye, A. N., & Zachmann, G. (2011). *Assessing the impact of EU ETS using firm level data* (Bruegel Working Paper 2011/08).

Anger, A., & Kohler, J. (2010). Including aviation emissions in the EU ETS: Much ado about nothing? A review. *Transport Policy, 17*, 38–46.

Blondiau, T., Delhaye, E., Proost, S., & Adler, N. (2016). ACCHANGE: Building economic models to analyse the performance of air navigation service providers. *Journal of Air Transport Management, 56*, 19–27.

Boeing. (2017). *Current Market Outlook 2017–2036*.

Coase, R. (1960). The problem of social cost. *The Journal of Law and Economics, 3*, 1–44.

Coase, R. (1992). The institutional structure of production. *American Economic Review, 82*(4), 713–719.

Council Directive 96/62/EC of 27 September 1996 on ambient air quality assessment and management.

Court of Justice of the European Union. (2011). *The directive including aviation activities in the EU's emissions trading scheme is valid*. Luxembourg: CJEU.

Decision 377/2013/EU of the European Parliament and of the Council of 24 April 2013 derogating temporarily from Directive 2003/87/EC establishing a scheme for greenhouse gas emission allowance trading within the Community.

Directive 2008/101/EC on amending Directive 2003/87/EC so as to include aviation activities in the scheme for greenhouse gas emission allowance trading within the Community.

EASA, EEA, & EUROCONTROL. (2016). *European Aviation Environmental Report 2016*.

Eden, C., & Ackermann, F. (1998). *Making strategy: The journey of strategic management*. London: Sage.

Efthymiou, M., Arvanitis, P., & Papatheodorou, A. (2016). Institutional changes and dynamics in the European aviation sector: Implications for tourism. In N. Pappas & I. Bregoli (Eds.), *Global dynamics in travel, tourism and hospitality* (pp. 41–57). Hershey Pennsylvania: IGI Global.

Efthymiou, M., Njoya, E. T., Lo, P. L., Papatheodorou, A., & Randall, D. (2019). The impact of delays on customers' satisfaction: An empirical analysis of the British Airways on-time performance at Heathrow Airport. *Journal of Aerospace Technology and Management, 11*. https://doi.org/10.5028/jatm.v11.977.

Efthymiou, M., & Papatheodorou, A. (2018). Environmental considerations in the Single European Sky: A Delphi approach. *Transportation Research Part A: Policy and Practice, 118*, 556–566.

Embraer. (2015). *Boeing e Embraer inauguram centro de pesquisa em biocombustiveis no Brasil* (press release). Available from https://embraer.com/global/en/news?slug=2333-boeing-e-embraer-inauguram-centro-de-pesquisa-em-biocombustiveis-no-brasil. Accessed 11 August 2019.

European Commission. (2012). *Single European Sky—Frequently asked questions.* Brussels: European Commission. Available from http://europa.eu/rapid/press-release_MEMO-12-934_en.htm. Accessed 11 May 2017.

European Commission. (2016). *Structural reform of the EU ETS.* Available from http://ec.europa.eu/clima/policies/ets/reform/index_en.htm. Accessed 1 July 2016.

European Commission. (2017). *Single European Sky.* Brussels: European Commission. Available from http://ec.europa.eu/transport/modes/air/single_european_sky_hu. Accessed 11 May 2017.

European Commission. (2018). *Reducing emissions from aviation.* Available from https://ec.europa.eu/clima/policies/transport/aviation_en. Accessed 13 July 2018.

Farmaki, A., & Papatheodorou, A. (2015). Stakeholder perceptions of the role of low-cost carriers in insular tourism destinations: The case of Cyprus. *Tourism Planning and Development, 12*(4), 412–432.

Freeman, R. E., Harrison, J. S., Wicks, A. C., Parmar, B., & de Colle, S. (2010). *Stakeholder theory: The state of the art.* Cambridge: Cambridge University Press.

Friedman, M. (1962). *Capitalism and freedom.* Chicago: University of Chicago Press and Phoenix Books.

Guastalla, G. (2014). *Performance indicator—Horizontal flight efficiency.* Brussels: Eurocontrol/PRU.

Havel, B. F., & Sanchez, G. S. (2012). Toward an international aviation emissions agreement. *Harvard Environmental Law Review, 36*, 351–385.

Holden, A. (2018). Environmental ethics for tourism—The state of the art. *Tourism Review, 74*(3), 694–703. https://doi.org/10.1108/TR-03-2017-0066.

ICAO Doc 10018. (2013). *Report of the assessment of Market-based Measures.* Available from http://www.icao.int/Meetings/GLADs-2015/Documents/10018_cons_en.pdf. Accessed 12 July 2016.

IATA. (2013). *The cost of the EU emissions trading scheme.* Montreal: IATA. Available from https://www.iata.org/publications/economics/Documents/EUETS-cost-briefing-march-2013.pdf. Accessed 28 July 2015.

IATA, EUROCONTROL, & CANSO. (2008). *Flight efficiency plan: Fuel and emissions savings.* Available from http://www.EUROCONTROL.int/sites/default/files/article/content/documents/nm/reports/flight-efficiency-plan.pdf. Accessed 6 October 2016.

ICAO. (2016). *Resolution A39-3 (2016): Consolidated statement of continuing ICAO policies and practices related to environmental protection—Global Market-based Measure (MBM) scheme.*

ICAO. (2017). *Why were international aviation emissions not included in the Paris Agreement at COP 21?* Available from https://www.icao.int/environmental-protection/Pages/A39_CORSIA_FAQ1.aspx. Accessed 13 July 2018.

Ihlen, O., & Berntzen, O. (2007). When lobbying backfires: Balancing lobby efforts with insights from stakeholder theory. *Journal of Communication Management, 11*(3), 235–246.

Jensen, M. C., & Meckling, W. (1976). Theory of the firm: Managerial behavior, agency costs and capital structure. *Journal of Financial Economics, 3*(4), 305–360.

Johnson, G., Scholes, K., & Whittington, R. (2010). *Exploring corporate strategy: Texts and cases* (9th ed.). Harlow: Financial Times Prentice-Hall.

Kantareva, M., Angelova, A., Iliev, L., & Efthymiou, M. (2015). *ICAO action plan on emissions reduction: Republic of Bulgaria.*

Korteland, M., & Faber, J. (2013). *Estimated revenues of VAT and fuel tax on aviation.* Delft: Delft CE.

Leggett, J., Elias, B., & Shedd, D. T. (2012). *Aviation and the European Union's Emissions Trading Scheme* (Congressional Research Service Report R42392). Washington, DC.

Liasidou, S. (2017). Drafting a realistic tourism policy: The airlines' strategic influence. *Tourism Review, 72*(1), 28–44.

Meleo, L., Nava, C. R., & Pozzi, C. (2016). Aviation and the costs of the European Emission Trading Scheme: The case of Italy. *Energy Policy, 88,* 138–147.

Moloney, K. (2006). *Rethinking public relations: PR propaganda and democracy* (2nd ed.). London: Routledge.

Niemeier, H. M., & Németh, A. (2012). Airline Mergers in Europe: An overview on the market definition of the EU commission. *Journal of Air Transport Management, 22,* 45–52.

Papatheodorou, A., & Lei, Z. (2006). Leisure travel in Europe and airline business models: A study of regional airports in Great Britain. *Journal of Air Transport Management, 12*(1), 47–52.

Pigman, G. A. (2007). *The World Economic Forum: A multi stakeholder approach to global governance.* New York: Routledge.

Porter, M. (1985). *Competitive advantage.* New York: Free Press.

PRB. (2016). *PRB White Paper: RP3 performance objectives, version 3.6.* Available from http://www.eusinglesky.eu/Documents/PRB%20Reports/Reference%20Period%20Three/RP3%20White%20Paper.pdf. Accessed 4 October 2016.

Preston, H., Lee, D. S., & Hooper, P. D. (2012). The inclusion of the aviation sector within the European Union's Emissions Trading Scheme: What are the prospects for a more sustainable aviation industry? *Environmental Development, 2,* 48–56.

Regulation (EC) No 1070/2009 of the European Parliament and of the Council of 21 October 2009 amending Regulations (EC) No 549/2004, (EC) No

550/2004, (EC) No 551/2004 and (EC) No 552/2004 in order to improve the performance and sustainability of the European aviation system.
Santos, G. S., de Arantes Gomes, R., & dos Santos, E. A. (2017). PPGIS as an urban planning tool around airports. *Journal of Air Transport Management, 69*, 269–278.
Saunders, M. N., Lewis, P., & Thornhill, A. (2015). *Research methods for business students*. Harlow: Pearson.
Scheelhaase, J., Maertens, S., Grimme, W., & Jung, M. (2018). EU ETS versus CORSIA–A critical assessment of two approaches to limit air transport's CO_2 emissions by market-based measures. *Journal of Air Transport Management, 67*, 55–62.
Schleich, J., Ehrhart, K. M., Hoppe, C., & Seifert, S. (2006). Generous allocation and ban on banking. In R. Antes, B. Hansjurgens, & P. Letmathe (Eds.), *Emissions trading and business*. Heidelberg: Springer.
Stabler, M. J., Papatheodorou, A., & Sinclair, T. M. (2010). *The economics of tourism* (2nd ed.). London: Routledge.
Stavins, R. N. (1995). Transaction costs and tradeable permits. *Journal of Environmental Economics and Management, 29*(2), 133–148
Stoney, C., & Winstanley, D. (2001). Stakeholding: Confusion or utopia? Mapping the conceptual terrain. *Journal of Management Studies, 38*(5), 603–626.
Tietenberg, T. H. (2010). *Environmental and natural resource economics* (10th ed.). New York: HarperCollins.
Williamson, O. E. (1975). *Markets and hierarchies: Analysis and antitrust implications*. New York: Free Press.
Williamson, O. E. (1979). Transaction-cost economics: The governance of contractual relations. *The Journal of Law and Economics, 22*(2), 233–261.
Williamson, O. E. (1998). Transaction cost economics: How it works; where it is headed. *De Economist, 146*(1), 23–58.

CHAPTER 7

Market-Based Measures: The European Union Emission Trading Scheme and the Carbon Offsetting and Reduction Scheme for International Aviation

Angela Stefania Bergantino and Luisa Loiacono

1 Introduction

Aviation is the fastest-growing source of pollutants. It is expected to increase its emissions by 300% by 2050, compared to 2005 levels (IATA 2015). More broadly, transport sector growth is a significant concern, in fact, from 1990 to 2007 in the European Union (EU), greenhouse gas (GHG) emissions from non-transport sectors fell 15%, while transport emissions increased by 33% (European Commission Climate Action

A. S. Bergantino (✉)
Department of Economics,
Management and Business Law, University of Bari, Bari, Italy

L. Loiacono
Department of Economics and Management,
University of Ferrara, Ferrara, Italy

T. Walker et al. (eds.), *Sustainable Aviation*,
https://doi.org/10.1007/978-3-030-28661-3_7

2018). It is clear that there is an urgent need to reduce carbon dioxide emissions from airplanes.

The European Union, as per the European Council conclusion 169/14 (EUCO 169/14, 2014), committed itself to reduce greenhouse gas emissions by 20% compared to 1990 levels between the years 2013–2020 and by 40 percent compared to 1990 within 2030. The purpose is to reduce GHG emissions in a "*cost-effective and economically efficient*" way.

In order to reach this goal, within the 2003 frameworks of the Kyoto Protocol and Horizon 2020, the European Union put in place the Emission Trading System (EU ETS) covering both 11,000 installations and (from 2012 on) more than 600 aircraft operators in 28 Member States of the European Union and additionally Iceland, Lichtenstein, and Norway.

The EU ETS, introduced with the Directive 2003/87/CE, also known as the ETS Directive, is a "cap and trade" scheme where a cap on total emissions allowed is set; in this system, one ton of CO_2 equivalent corresponds to one allowance. An important implication of the EU ETS the companies covered has to measure and report their carbon emissions annually.

2 The European Union Emission Trading System

The EU ETS Handbook from the European Commission (European Commission Climate Action 2015) explains both the general and the Aviation Emission Trading System.

The EU ETS is divided into 4 phases:

1. Phase 1 (from 2005 to 2007): "Learning by doing" phase, the number of allowances turned out to be excessive, consequently the price of first-period allowances fell to 1 € per allowance in 2007.
2. Phase 2 (from 2008 to 2012): Characterized by the surplus of unused allowances and credits because of economic downturn, aviation was brought into the system in 2012.
3. Phase 3 (from 2013 to 2020): The EU-wide cap on emissions, in the general ETS, fell by 1.74% each year (Linear Reduction Factor), and there was a progressive shift towards auctioning of allowances in place of cost-free allocation.
4. Phase 4 (from 2021 to 2030): The Linear Reduction Factor will be 2.2% each year from 2021 to 2030.

The whole EU ETS caps 50% of total EU GHG emissions. The "cap and trade" system allows trading of emission allowances so that the total emissions of the installations and aircraft operators stay within the cap, this encourages the use of least-cost measures to reduce emissions.

The EU ETS covers:

- carbon dioxide (CO_2) from:
  - power and heat generation;
  - energy-intensive industry sectors including oil refineries, steelworks and production of iron, aluminum, metals, cement, lime, glass, ceramics, pulp, paper, cardboard, acids, and bulk organic chemicals;
  - civil aviation;
- nitrous oxide (N_2O) emissions from all nitric, adipic, and glyoxylic acid production;
- poly fluorinated chemicals (PFC) emissions from aluminum production.

In the scheme, the maximum quantity of the aforementioned emissions is set. The carbon price is not given, it is set by trading and determined by the balance of supply and demand. In the scheme, only a certain amount of allowances is traded, while a relevant percentage is given for free to the operators.

Giving the allowances free of cost is also called "grandfathering," the reasons why allowances are freely assigned are both to avoid carbon leakage and to avoid market distortions; however, after 2020, grandfathering will become the exception and auctioning the general rule. A relevant amount of allowances given for free distorts the market, as only a small percentage of allowances are actually traded. This creates a narrow market as operators do not need to buy the allowances for the CO_2 equivalent emitted therefore resulting in a lower cost.

In 2013, in the general EU ETS scheme, there was an over-allocation of allowances, this was due both to an external factor, the economic slowdown, and an internal factor, over-allocation of emission permits. The European Parliament, in order to ensure the functioning of the market, stated that in such exceptional circumstances, there was the option of adjusting the market by adapting the timing of a maximum of

900 million allowances. The initiative, called "backloading",[1] consisted in moving a set amount of allowances from the period 2013–2015 to 2019–2020. After the "backloading" initiative, a new mechanism, created to face market disturbances, was designed: the Market Stability Reserve (MSR) (DECISION (EU) 2015/1814). The MSR, established in 2018, started operating from 2019, and the 900,000 allowances to be reintroduced in the market were held in it. Allowances from the "backloading" initiative and the ones not allocated from the new entrant reserve were held in the MSR. This measure made the market more efficient as it created scarcity of allowances (in fact, as the scarcity pricing theory suggests, the equilibrium price is reached when there is no surplus, either on the supply or on the demand side).

2.1 The Aviation European Union Emission Trading System

Directive 2008/101/EC amended the aforementioned Directive 2003/87/EC by including aviation activities in the Emission Trading Scheme (ETS). It was included during the second phase of the scheme in 2012. The civil aviation, in fact, accounts for 13.1% of the total transport emissions in Europe, second only to the road transport (accounting for 72.8%) (European Parliament 2018). The rationale behind the inclusion of aviation in the scheme is to limit the increase in global temperature to less than 2 °C above "pre-industrial" levels. The Directive 2008/101/EC establishes the cap at 97% of historical emissions in 2012 and at 95% between 2013 and 2020.

In the aviation sector, the amount of free allowances assigned to each operator is based on a benchmark established in 2011 by the European Commission and the European Economic Area (EEA) Joint Committee.

In order to understand how the cap is set and how the amount of free allowances in allocation is calculated, the necessary steps are here reported.

The steps to calculate the benchmark:

$Total\ Allowances_{2012} = 0.97 \times Historical\ Emissions_{2004\text{–}2006}$

$Free\ Allowances_{2012} = 0.85 \times Total\ Allowances_{2012}$

$Benchmark_{2012} = Free\ Allowances_{2012}/Reported\ RTK_{2010}$,

where RTK stands for Revenue Ton-Kilometers.

[1]Decision 1359/2013/EU.

RTK is a standard measurement used to quantify the revenue generated by passengers and freight carried, taking into account the distance flown. According to EUROCONTROL, one RTK is generated when one metric ton of revenue load is carried for one kilometer. Directive 2008/101/EC states that when calculating RTKs the following standard measures have to be respected: passenger weight of 100 kilos and 95 kilometers in addition to the Great Circle Distance (GCD, a standard distance between two airports) of each flight. As mentioned before, the amount of free allowances in allocation is conspicuous, thus it distorts the auctioning system.

The baseline for historical emissions[2] is derived from the European Organization for the Safety of Air Navigation[3] and from actual fuel consumption data provided by aircraft operators (European Commission Climate Action 2015).

The historical emissions considered are the mean of the emissions in 2004 (209,123,585 tons of carbon dioxide), 2005 (220,703,341 tons of carbon dioxide) and 2006 (228,602,103 tons of carbon dioxide), resulting in 219,476,343 tons of carbon dioxide. These baselines do not change over the years, while the sector itself is dramatically growing over the years.

Eighty-two percent of this amount is issued to operators free of charge, 15% is auctioned and 3% is set aside in a "special reserve" to cover the free allocations to new entrants and fast growers (companies growing at an average rate of 18% annually, between 2010 and 2014).

In its initial goal, the EU ETS covered all intra-EEA and extra EAA flights. The rationale behind this choice was to avoid carbon leakage. However, the EU ETS has limited its scope to intra-EEA flights since 2013. Originally, the scheme was validated by the Chicago Convention, which recognized expressly the right of each Contracting Party to apply on a non-discriminatory basis its own laws and regulations (Brinke and Faber 2011).

[2]Historical emissions: Average of the annual emissions in years 2004, 2005, and 2006 from aircraft performing activities covered by the ETS.

[3]European Organization for the Safety of Air Navigation, also known as EUROCONTROL, collected information from (a) fuel information from Aircraft Operators; (b) databases from the Central Route Charges Office and the Central Flow Management Unit.

This limitation, called "stopping the clock," was driven by diplomatic objections from countries (such as China, India, and USA), which argued that the EU ETS had no competence on extra EU countries. The European Union was accused of breaking international law by applying the scheme to distances traveled outside the EU, where the EU had no jurisdiction. The adjustment of free allocation to aircraft operators from the full scope to the reduced scope went through the Regulation 421/2014, which introduced Article 28A as an amendment of Directive 2003/87/C. The limitation of the scope significantly changed the layout of the scheme, implicating two main consequences:

- The system had, overall, a much different impact on emissions covering as it excluded an enormous part of the civil aviation sector: the routes outside the EEA, which are the longest ones, were completely excluded.
- The system went through a difficult phase, and it had to stop the auctioning system for one year, thus distorting the market even more: risk and uncertainty are typical market distortions which are usually solved by the introduction of new regulation.

After the "stopping the clock provision," the EU ETS limited its scope to the flights within the EEA–28 EU Member States plus 3 European Free Trade Association countries: Austria, Belgium, Bulgaria, Croatia, Cyprus, Czech Republic, Denmark, Estonia, Finland, France, Germany, Greece, Hungary, Iceland, Ireland, Italy, Latvia, Liechtenstein, Lithuania, Luxembourg, Malta, the Netherlands, Norway, Poland, Portugal, Romania, Slovakia, Slovenia, Spain, Sweden, and the UK. Flights beyond the EAA will be excluded, including those to and from: Spain's Canary Islands; the French territories of French Guiana, Guadeloupe, Martinique, Reunion, and Saint Martin; and the Portuguese Azores and Madeira.

The Aviation Emission Trading System includes some exemptions:

- air transport operators operating, for three consecutive four-month periods, fewer than 243 flights per period;
- flights with overall CO_2 emissions lower than 10,000 tons per year;
- flights performed in the framework of public service obligations (PSO) on routes within remote regions or on PSO routes where the capacity offered does not exceed 30,000 seats per year. PSOs

take into account social, environmental and regional development, in air transport, example of routes covered are: Dublin to Kerry and Donegal, Islands to continental Greece, Sardinia and Elba Islands to continental Italy, Corsica Island to continental France;
- military flights;
- search and rescue flights;
- humanitarian flights;
- aircraft with a take-off mass of less than 5700 kg.

The funds raised through auctions will go to Member States, which also have the role of determining airlines free allocations and supervise their compliance.

The airlines are assigned to each Member State based on two different criteria:

- if the operator is based in the EU, the airline is assigned to the Member State that issued its license;
- if the operator is based outside the EU, it is assigned to the Member State to which it attributes the greatest amount of emissions (European Commission Climate Action 2015).

Directive 2003/87/EC states that, in general, Member States should spend 50 percent of the revenues from auctioning:

- on reducing climate change in the European Union and in third countries;
- on developing renewable energies and storage;
- on funding research and development.

The directive states that auction revenues from aviation should be used, in particular, in the fields of aeronautics and air transport, to reduce emissions through low-emission transport and to cover the cost of administering the Community scheme. More broadly the EU ETS Directive states that at least 50% of the revenues from the auctioning of general allowances should be used to finance climate and energy activities: A European Union analysis of the use of auction revenues by the Member States show that, over the years 2013–2015, at least 80% of the revenues were used for climate and energy purposes (Directorate-General for Climate 2017).

Each calendar year, operators have to monitor and report their emissions in the Union Registry by 30th of April of the following year. Then, the confirmation of surrendered allowances is up to Member States' national administrators. If an operator does not surrender allowances, it has to pay a fine for each excess ton of GHG emitted.

Every year, aircraft operators have to report the annual Monitoring Plan. Verification has to be carried out by accredited verifiers.

In the EU ETS, participants, until 2012, could use international credit from Kyoto Protocol (in place until 2020):

- Clean Development Mechanisms: An industrialized country with a greenhouse gas reduction commitment (Annex I countries) implements emission reduction projects in developing countries. These projects can earn tradable Certified Emission Reductions (CERs), and each certificate corresponds to one ton of carbon dioxide (United Nations Climate Change 2018).
- Joint Implementation: Industrialized countries can purchase Emission Reduction Units (ERUs) from projects that reduce emissions in other industrialized countries.

During Phase 3, CERs and ERUs are no longer direct compliance units but have to be converted into EU allowances, thus the eligible international credits have to be exchanged for allowances. This change guaranteed a more integrated market; in fact, it created a more transparent and efficient system as the airlines were obliged to buy the allowances from the auctioning market and could not exceed the threshold by using other instruments such as CERs and ERUs.

For an international credit to be eligible for the EU ETS, the project does not have to be:

- nuclear projects;
- reforestation activities (LULUCF);
- projects involving the destruction of industrial gases;
- CERs from projects starting after 2012 unless the project takes place in the least developed countries (LDCs).

These restrictions are qualitative standard that has to be respected and they constitute a negative list of projects.

For the period from 2013 to 2020, a limit of 1.5% of verified emissions to use for international credits has been set for each aircraft operator.

However, in two years, there will be a new scheme in place, the global Market-based Measure (MBM) introduced by the International Civil Aviation Organization (ICAO), called the Carbon Offsetting and Reduction Scheme for International Aviation (CORSIA). CORSIA is an offsetting mechanism, which enables aviation growth as far as emissions are compensated through international offset credits (mainly generated in non-aviation sectors).

Participation in CORSIA will be on a voluntary basis from 2021 to 2026 and will become compulsory from 2027 on (yet, CORSIA will exclude the least developed countries and landlocked developing countries). Because ICAO has a no competence agreement for domestic aviation, the upcoming scheme will not include domestic flights. An important consequence of this feature is that the flights within the countries will not be covered, thus there may be the requirement for national authorities to rule the domestic flights: This non-harmonized ruling would probably distort both the CORSIA and the domestic markets as there would be two different systems with different scopes, rules and goals. Moreover, there is the risk that the EU ETS will be phased out by the introduction of the CORSIA (IATA 2017) since there may be an overlapping jurisdiction, thus lowering the overall cap since the CORSIA mechanism is much less ambitious from an environmental perspective.

3 The European Energy Exchange and the Intercontinental Commodity Exchange

Auctioning is considered the general rule to allocate allowances to the operators in the EU Emission Trading System, and the auctioning system reinforces the "polluter pays" principle (Art. 19 of the Treaty on the Functioning of the European Union). The carbon price indicates, given demand and supply, the EU ETS participants' willingness to pay. One allowance equals one ton of carbon dioxide. The European Commission (EC) regulates the auctions. When the market was created, the aim was to create a common EU platform. However, the EC gave member states the option to opt out and set up a national platform (UK, Poland and Germany chose to opt out). The established common platform is the European Energy Exchange and it auctions allowances on behalf of Member States.

In the auctions, there are two different kinds of allowances:

- European Union Aviation Allowances (EUAAs): Their use is limited to aircraft operators;
- European Union Allowances (EUAs): They are used by all operators and subjected to the ETS (including aircraft operators).

There are two different platforms managing the exchange:

- the European Energy Exchange (EEX) in Leipzig is the common platform used by 25 Member States plus the EEA-EFTA States. It also hosts Germany's and Poland's platforms;
- the ICE Future Exchange in London, where the UK's opt-out platform takes place.

However, all platforms must give all eligible bidders equal access to auctions and no preference can be given to companies registered in any particular Member State.

The auctioning, ruled by the EU ETS Auctioning Regulation, takes place in a harmonized, transparent and non-discriminatory way. It is cost-efficient and it guarantees full, fair, and equitable access to auctions in order to minimize the risk of market abuse.

The auction system works as a single-round, sealed bid (where all bidders simultaneously submit bids and no bidder knows what the other participants' bid) and the closing price is uniform.

All successful bidders pay the same clearing price. The successful bidders are the ones who placed bids at or above the clearing price. During the bidding window, bidders place bids specifying the number of allowances and the price. Once the window is closed (after at least two hours), the platform publishes the clearing price set where demand equals supply. All successful bidders pay the same price, regardless of the price they specified in their bids. The amount of allowances for sale is set before the bidding.

Table 1 shows yearly data: the average price, the number of allowances traded, the average number of participating and successful bidders and the total revenue (average price times total allowances traded).

Table 1 Aviation allowances descriptive data (including the German, Polish, and British markets)

Year	*Average price*	*Total allowances traded*	*Average bidders*	*Average successful bidders*	*Total revenue*
2012	€7.01	2,500,000	6.0	5.0	€17,525,000
2013	€0.00	0	0.0	0.0	€0
2014	€5.84	9,278,000	13.0	8.0	€54,183,520
2015	€7.26	16,390,500	11.2	6.4	€118,995,030
2016	€5.31	5,997,500	10.3	4.8	€31,846,725
2017	€7.24	4,730,500	6.7	5.5	€34,248,820
2018	€17.96	5,601,500	7.7	5.3	€100,602,940

Source European Energy Exchange and Intercontinental Exchange

As shown in Table 1, the number of bidders is very low (5.3 in 2018) meaning that the market is mainly participated by intermediaries (Heindl 2012). The low number of bidders favors collusive behavior which does not favor the market equilibrium.

However, since 2018, a relevant reform to the auctioning market was introduced. The auctioned products were not considered financial instruments before the directive on the markets of financial instruments 2014/65. Since January 2018, the emission allowances have been classified as financial instruments. This reform allowed other agents to invest in the medium and the long term thus increasing the number of participants in the market.

The Auctioning Regulation states that allowances have to be auctioned as spot products and delivered within five working days.

However, other than the primary market, the auctioned allowances are traded also in the secondary market.

The eligible bidders before the reform were:

- ETS operators (and their subsidiary, parent or affiliate);
- business groupings with legal identity, e.g., European Economic Interest Grouping (EEIG);
- government entities controlling EU ETS operators;
- investment firms and credit institutions under EU legislation.

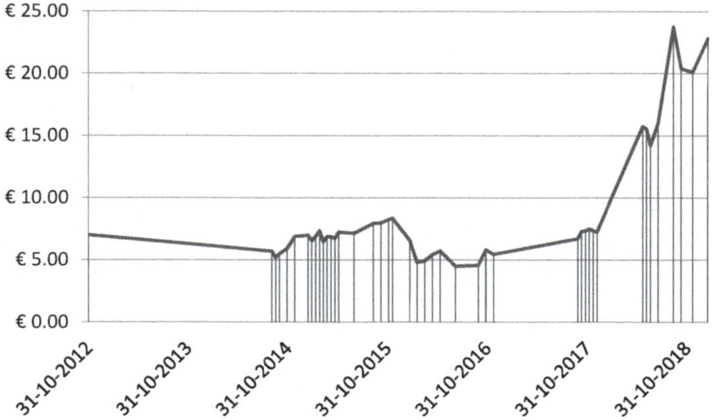

Fig. 1 Daily closing EUAA prices: EU-wide platform, plus Poland, Germany, and the UK (*Source* European Energy Exchange and Intercontinental Exchange)

Intermediaries and EEIGs offer small emitters and small and medium enterprises to minimize transaction costs and better managing the allowances purchased. Eligible bidders, in order to participate in the auctions, have to pay fixed and variable fees. Each year, the European Commission publishes the auction calendars specifying the number of allowances per auction and the dates of the auctions. In 2016, EEX was reconfirmed as the common EU platform (the Auctioning Regulation (Directive 2010/1031/EU) states that the platform has to be appointed for maximum five years).

The price trend of the European Aviation Allowances closing prices shown in Fig. 1 and did not vary significantly until 2018. It fluctuated in a pretty constant way from 2012 to October 2015. Then, in November 2015, the price dropped and the trend remained slightly lower until 2018 when the price rose to 15 €. From Table 1, it emerges that the price does not depend on the quantity of allowances auctioned; however, it must be reminded that the price is driven by the general allowances market; therefore, the comparison between the two markets (Fig. 2) is fundamental.

There is a significant difference in the frequency of auctions (Fig. 1) during the years:

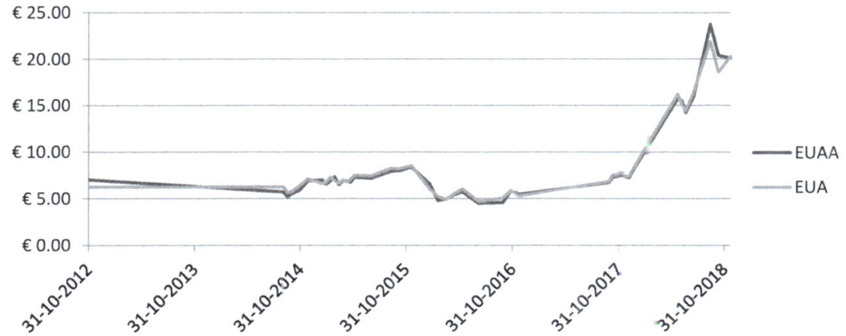

Fig. 2 EUAA prices compared to EUAs prices: EU-wide platform, Poland, Germany, and the UK (*Source* European Energy Exchange and Intercontinental Exchange)

- In 2012, there was one only auction in January (mainly because of the "stopping the clock" provision).
- In 2014, there were five late auctions (from September to November).
- In 2015, there were two auctions per month from January to April and six more in the remaining eight months.
- In 2016, there was approximately one auction per month, nine auctions in total from January to November.
- In 2017, the auctions did not start until September, and the year closed with six auctions.
- In 2018, the auctions were seven, from May to November, and the price was significantly higher.

The number of yearly auctions strongly affected the airlines behavior since the market was characterized by uncertainty, and it was strongly connected to the EUAs market.

Comparing the two allowances available in the auction: The EUAAs (limited to aircraft operators) with the EUAs (used by all operations and subjected to the ETS) is significant since aircraft operators can purchase allowances from the general market[4] (the opposite is not possible).

[4]The opportunity for aircraft operators to buy allowances from both the general and the aviation market, to a certain extent, limits the EU ETS's objective of creating scarcity of allowances (DEHSt 2018).

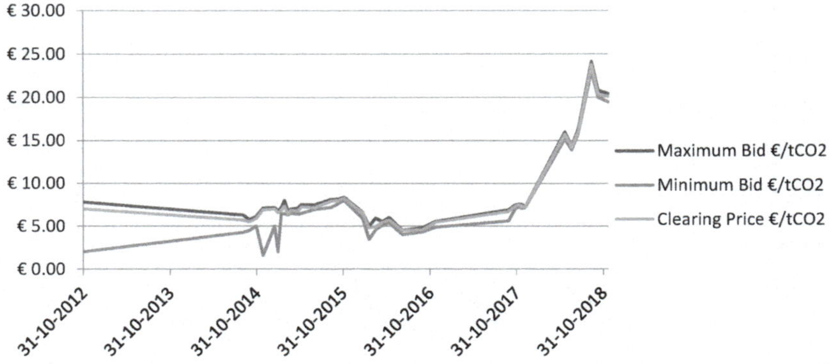

Fig. 3 Minimum, maximum and clearing price for EUAAs (*Source* European Energy Exchange and Intercontinental Exchange)

As shown in Fig. 2, the trend is similar, and there is an almost perfect correspondence between the price of the European Union Allowances (EUAs) and the price of the EUAAs. The implication is that the EUAAs market is strongly dependent on the EUAs market because a shock in prices would lead airlines to abandon the aviation market. Moreover, the possibility for airlines to purchase from the general market makes the cap weak: The airlines can exceed the cap by purchasing from allowances the EUAs market.

The EUAAs prices examined in Fig. 3 also consider the clearing price related to the minimum and maximum bids.

After the auction closes, all the successful bidders pay the same price for the allowances they bid for: The clearing price is always very close to the maximum price, while there is a wider spread with the minimum price (which sometimes reaches 2 € and 3 €). This figure indicates that the variation within the minimum and maximum price is very low and the closing price is very close to the minimum and maximum price, meaning that within the bidding window there is not fluctuation.

During the first phases, the general EU ETS scheme suffered from oversupply, which rendered the scheme ineffective and the prices very low. In 2017, in order to adjust the system, the European Union introduced the aforementioned MSR, which influenced both general and aviation prices since the two markets are very strongly related (the aircraft operators can purchase allowances from both the general market and the dedicated one). In fact, the prices follow parallel trends even if the reforms only affect the general ETS (Fig. 2). On November 2018, Arias Cañete, on behalf of the European Commission, declared that the increase in prices in 2018 was due to the strengthening of the MSR, which addresses the surplus of permits in the carbon market by removing surplus allowances from 2019 on. The carbon prices rose in anticipation of its impact. In fact, when the demand decreases, either demand must fall, or prices must rise consequently.

Thanks to the reforms, it is expected that over the years 2019–2023 the permits cut will be around 1.5 billion.

Thus, the high prices registered in 2018 are mainly due to the MSR and the LRF. The MSR does not only include 900 million allowances suspended between 2014 and 2016, but also acts as a market regulator in the long term.

From January 2019 on, the aforementioned MSR, which will create a healthy scarcity in the market, will follow these rules:

- Rules for the retention of rights: If the published number of emission rights in circulation exceeds 833 million tons, the MSR will cancel 24% of the surplus each year up to 2023 and 12% thereafter.
- Rules for the release of rights: If the published number of emission rights in circulation is less than 400 million tons, 100 million rights will be released in scheduled auctions. If the amount of permits in the reserve is less than 100 million tons, all permits will be released.

4 The European Union Transaction Log

Commission Regulation 748/2009/EU, amended by Commission Regulation 82/2010/EU, provides the "list of aircraft operators which performed an aviation activity listed in Annex I to Directive 2003/87/EC on or after 1 January 2006 specifying the administering Member State for each aircraft operator."

The European Union Transaction Log (EUTL) is a registry covering all 31 countries participating in the EU ETS created in 2012. It holds both stationary installations and aircraft operators, and it is managed by the Directorate-General for Environment.

The registry collects:

- companies' accounts;
- allowances' transactions;
- annual verified carbon dioxide emissions;
- surrendered emissions;
- sources of surrendered emissions.

In the registry, all the installations and airlines covered by the scheme are listed together with the accounts' details. The annual verified carbon dioxide emissions and the allowances in allocation are reported in the registry.[5] Account holders can choose to "bank" allowances from one year to another and from one phase to another (this was excluded in the first phase, when airlines where less willing to buy allowances because they could not "bank" them to the following phases); furthermore, the allowances can be either used or sold. Deadlines for registration of verified emissions and surrendered units are, respectively, in March and in April of the following year. Account holders can be installation operators, airline operators or traders. The registry traces all the transactions and exchanges for each operator, it marks the exchange of the general and aviation allowances, the Certified Emission Reductions, and the ERUs. The EUTL is not an exchange platform; in fact, the auctions take place in the European Energy Exchange and in the Intercontinental Commodity Exchange (the UK only). All the operations in the registry are authorized and traced by the central Log of the European Union.

In the section, "Allocation to Aircraft Operators" of the Registry 673 airlines assigned to 31 countries are represented, the airlines are covered in both the second and the third phase. Originally, the scheme covered 900 aircraft operators since it also included flights leaving the EU28 plus Iceland, Liechtenstein, and Norway.

The first six countries, in terms of number of airlines assigned, are: The UK (158 operators, 23%), Germany (113 operators, 17%), France (91 operators, 14%), Spain (50 operators, 7%), Italy, and Ireland (41 operators and 6% each). The position of the UK is due both to domestic

[5] Each account refers to its national administrator—some national administrators are: "Istituto Superiore per la Protezione e per la Ricerca Ambientale" in Italy, "Climate Registry" in Belgium, "Registre française des quotas et credits d'émission de GES" in France.

Table 2 All countries' aviation emissions (including companies excluded after 2012)

Year	*Freely allocated allowances*	*Verified emissions*
2012	173,817,206	83,978,299
2013	32,133,577	53,484,229
2014	32,069,612	54,829,920
2015	31,713,666	57,087,006
2016	31,587,393	61,453,056
2017	32,659,446	64,246,416

Source European Environment Agency

airlines and to the London Heathrow airport, which is the main hub for international flights in Europe. Thus, many non-EU-based airlines are assigned to the UK.

From Table 2, it emerges that verified emissions follow an increasing trend over the years (with the exclusion of 2012, when also the flights from and to the area covered by the system were covered). On the other hand, the freely allocated allowances are constant with a slight increase in 2017 due to new entrants and fast growers. This means that airlines have to buy more allowances from the auctioning market, thus making the system more efficient.

The passage from 2012 to 2013 presents a large difference compared to the transitions in the following years. This is due to two main reasons:

- Allowances in allocation decreased because of the "stopping the clock" provision, between end of 2012 and 2013, which limited the ETS to cover the flights operated only within the EEA plus Iceland, Liechtenstein and Norway.
- The carbon dioxide cap was reduced from 97% of average 2004–2006 emission levels in 2012, to 95% of 2004–2006 levels in 2013 then distributed in the following way: 82% granted for free to aircraft operators, 15% auctioned, and 3% held in a special reserve for fast-growing.

5 The Model

Focusing on the EU ETS and its impact at the firms' level, the existing literature mostly examines the general scheme covering more than 11,000 power stations and manufacturing plants. This analysis allows us to state whether the scheme is effective: Is the variation in emissions due to a variation in the turnover or to the scheme? In other words, does the introduction of the scheme affects the companies' behavior?

The studies considered here focus on the effectiveness of the ETS (Abrell et al. 2011).

With the aviation scheme created as a segment of the general EU ETS scheme, the two systems are very similar in terms of design and legislative framework, thus the investigations conducted on the installations are consistent with the analysis of the impact on the aviation sector.

The analysis conducted is based on airlines' emissions of CO_2, performance indicators, and production inputs for the period 2013–2016.

The panel data regression analyzes the firms' data impact on verified emissions and the behavior of firms in terms of verified emissions over the years.

A particular focus is necessary on the first three years (2012–2014). In 2012, the scheme covered all the flights from and to the EU28 (plus Iceland, Lichtenstein, and Norway), but the "stopping the clock" provision limited the Emission Trading System to cover only the flights operated within the EEA plus Iceland, Liechtenstein, and Norway. In 2013, because of the reduced scope, there were no auctions, thus companies did not have to pay for allowances before 2015.

The aviation industry analysis is based on data from various sources:

- Balance sheets from the database Amadeus. The collected data is turnover, number of employees, total assets, and cash flow.
- Average aviation allowances (EUAAs) price per year. The sources are EEX and ICE.
- Verified Emissions and Compliance from the European Union Transaction Log.

In order to study the overall trend, 57 airlines have been considered.[6] They account for the 58.2% of total verified emissions in the whole period from 2012 to 2016.

The companies considered do not include airlines based outside the 28 countries. The rationale behind this choice is that companies outside this territory do not have a consistent share of traffic covered by the scheme and the turnovers would be not comparable.

[6] Data is not available for all the companies for the whole period. In order for the panel data to be balanced, companies with lacking data were excluded.

The considered variables and the expected results are considered before the analysis is run.

- *Verified emissions* are the dependent variable. The purpose of this regression is to study what influences the verified emissions and to what extent they are influenced. Verified emissions are expressed in tons of CO_2 per year and per company.
- The *Fleet* variable considers all the aircraft purchased per year and per company; however, it does not consider the total fleet per year. The expected result can be either positive or negative. On the one hand, if the coefficient is positive, we assume that the purchase is due to the fleet growth and an increase in passenger carried; on the other hand, if the coefficient is negative or close to zero, we assume that the fleet purchased is due to fleet renewal and thus to aircraft substitution.
- The *Turnover* is expressed in euro and ranges from thousands to millions. Turnover is expressed per year and per company. The expectation is that the coefficient has a positive sign, as airlines' revenue is mainly based on ticket sales.
- The *Total Assets* variable indicates the material and immaterial assets. We expect the coefficient to be positive since an increase in the total assets should correspond to an increase in aircraft owned, thus the traffic is expected to grow.
- *Employees* indicate the number of employees per year and per company. We expect that when employees increase, verified emissions increase as well because it would imply an increase in distance flown.
  However, this result may be distorted by the characteristics of the civil aviation labor market. In fact, the aviation sector, especially in low-cost companies, self-employment and agencies are commonplace: It is estimated that in 2015 in low-cost companies only 52.6% had a direct employment contract. This practice implies that to an increase in traffic does not always correspond an increase in employees (pilots).
- *Compliance* refers to the surrendered emissions per company per year. Each March, companies surrender the number of allowances corresponding to the difference between verified emissions and allowances in allocation. The variable is binary; it is 0 when the airline is compliant and 1 when it is not compliant or when the airline

surrenders the allowances with delay. The coefficient indicates the change from 0 to 1. Since it is a dummy variable, the value of 0 compliance is omitted and works as a baseline. We expect that when airlines are not compliant, verified emissions increase.

- The *Price* is expressed in euro. It is the same for all the airlines and it changes over the years. In 2013, because of the "stopping the clock" provision, the value was zero and there were no EUAAs traded. In the following years, the price was calculated as the yearly average of EUAAs prices. As the price is set through auctions, we expect the coefficient to be negative, meaning that when the price increases, the verified emissions decrease.

The average flight distance in Europe is 2270 kilometers (European Commission, 2007), which is a short–medium-haul, with EasyJet and Ryanair flying an average sector length of 1105 and 1239 kilometers, respectively. For this reason, the distance flown is not considered in the analysis; short- and medium-haul flights are similar in terms of emissions since the peak is during take-off and landing. For example, flying from Rome to Milan (475 kilometers) corresponds to 90.8 kilos of carbon dioxide emissions per passenger, while flying from Rome to Berlin (1197 kilometers), accounts for 119.8 kilos of carbon dioxide per passenger: The emissions per passenger increase by 32% when distance increase by more than 150% (ICAO) (Table 3).

In order to understand the magnitude of the results, we consider as a baseline the average of the single variables (Table 4).

The mean of each year is calculated as follows: $\bar{x} = \frac{1}{j}\sum_{j=1}^{n} x_j$, where $j=1, \ldots, n$ (where n is the sample size). The mean reported in Table 4 is the average value for the years considered (2013–2016).

On average, an aircraft flying a short-haul[7] route from Rome to Milan (468 kilometers) and carrying 150 passengers would emit 12.13 tons of carbon dioxide (ICAO 2016). This corresponds approximately to 2.59 tons of carbon dioxide per 100 kilometers flown by an aircraft carrying 150 passengers.

[7]According to EUROCONTROL (2015): A flight is short-haul when the airport to airport distance is less than 1500 km; medium-haul when the airport to airport distance is less than 4000 km but more than 1500; long-haul flight when the airport to airport distance is more than 4000 km.

Table 3 Linear regression, correlated panels corrected standard errors (PCSEs)
$\ln \text{verified emissions} = \beta_0 + \beta_1 \textit{fleet} + \beta_2 \ln \textit{turnover} + \beta_3 \ln \textit{totalassets} + \beta_4 \textit{employees}$
$+ \beta_5 \textit{compliance} + \beta_6 \textit{price} + \varepsilon$

Group variable	id	No. of obs.	228
Time variable	Year	No. of groups	57
Panels (balanced)	Correlated (balanced)	Obs. per group	4
Autocorrelation	No autocorrelation		
Estimated covariances	1653	R^2	**0.5802**
Estimated autocorrelations	0	Wald → $\chi^2(8)$	3914.99
Estimated coefficients	7	Prob > χ^2	0
Panel corrected standard errors			
ln verified emissions		Coeff.	$P>\|z\|$
Fleet		0.08257	0.01
ln turnover		0.41258	0.00
ln total assets		0.23325	0.00
Employees		−0.00000	0.10
Compliance		0.53497	0.09
Price		−0.01676	0.08
Cons.		1.01778	0.09

"ln" stands for natural logarithm
"β_0, "β_1" until "β_6" are coefficients
"verified emissions, "fleet", "turnover", "totalassets", "employees", "compliance", "price" are variables as "x_1", "x_2", "x_3" and so on
"ε" is the error term

Table 4 Mean values of verified emissions, fleet, turnover, total assets, cash flow, number of employees, and price from 2013 to 2016

Variable	*Mean*	*Std. Dev.*
Verified emissions (tCO_2)	620,552.00	1,294,432.00
Fleet (# aircraft purchased)	3.24	5.00
Turnover (€)	2,070,000,000.00	5,740,000,000.00
Total assets (€)	1,970,000,000.00	5,740,000,000.00
Employees (# employees)	6,128.00	20,014.30
Price (€)	4.60	2.80

When analyzing the estimated coefficients resulting from the Prais-Winsten regression, the analysis shows that there is a positive relationship between Verified Emissions and Fleet, Turnover, and Total Assets meaning that the purchase is due to the fleet growth and an increase in passenger carried; and a negative relationship between verified emissions and employees and EUAAs price.

- R^2. The R^2 of the model is 58.02%, which is a satisfying value, meaning that 58% of the dependent variable variation can be explained by the model.
- *Fleet*. As there is an increase of verified emissions, the fleet purchased is more likely due to an increase of passengers carried. In fact, when one additional aircraft is purchased, verified emissions increase by about 8.63%.
- *Turnover*. The coefficient indicates that a one percent increase in turnover increases the expected verified emissions by 0.18%.
- *Total Assets*. The coefficient indicates that a ten percent increase in total assets also increases the expected verified emissions by 10.6%.
- *Employees*. The employees' coefficient is zero. The expectation was for the coefficient to be positive. This unexpected result may derive from the way the number of employees is calculated (e.g., not considering external employees).
- *Compliance*. The compliance is not significantly different from zero, thus the coefficient is excluded from the analysis of the results.
- *Price*. The coefficient of the average yearly price is negative. When the price increases by 1 €, the verified emissions decrease by 1.69%. Consequently, the price, ranging from a minimum of 0 € in 2013 to a maximum of 7.26 € in 2015, is effective because it has the desired impact on verified emissions: As the price increases, verified emissions decrease in a significant way.
- *Coefficient*. As the dummy variables are not significant, the intercept, which is the expected mean value of the verified emissions when all the regressors are zero, is not particularly relevant.

6 Conclusions

The Aviation EU ETS suffers from various critical issues:

- "Stopping the clock" provision;
- Non-liquidity;
- Excess of free allocations;
- Scarce number of bidders;
- Introduction of the CORSIA scheme.

The EUAAs market is a non-liquid market, thus the auction system is distorted. Moreover, the emission cap applied to the aviation sector is much wider than the cap applied to the installations.

In addition, a relevant concern is the introduction of the global CORSIA, which could phase out the EU ETS, replacing it with a much less ambitious system.

The purpose of this study was to analyze, at the firms' level, which factors impact the verified emissions in order to find out if and to what extent the introduction of the auctioning market decreases GHG emissions. The sample we used considered 57 airlines accounting for 58.2% of total verified emissions during the period 2012–2016. We find evidence that verified emissions decrease when the price increases. Yet, they also increase when fleet, turnover and total assets increase. This result shows that given the change in the distance flown (controlled with fleet, turnover and cash flow), the scheme is effective and it influences the airlines behavior in terms of verified emissions. Given the current increase in prices, the airlines will bear higher carbon dioxide costs in the future: This should push forward the effectiveness of the scheme, possibly making the negative *price* coefficient higher. The companies, in fact, in order to face the increase in distance flown, could limit their emissions by using new technologies, investing in less polluting aircraft, and adopting sustainable fuels.

In order to run this analysis, we analyze previous studies, which considered the general EU ETS, the most relevant is the one analyzing the effectiveness of the general ETS (Abrell et al. 2011).

From the analysis run, it emerges that the EU ETS scheme is giving the expected results as, when price increases, verified emissions decrease. However, despite this encouraging result, it emerges that from 2015 to 2016, while airlines overall turnover decreased by 5.7%, verified emissions increased by 7.3%.

As an important implication of the scheme, the "polluter pays" principle applied here is positively influencing the Member States investments in reducing climate change through low-emissions transport (Directive 2003/87/CE).

Moreover, from 2018 on, the allowances started a robust increase, thus making the market much more effective.

References

Abrell, J., Ndoye Faye, A., & Zachmann, G. (2011). *Assessing the impact of the EU ETS using firm level data* (Bruegel Working Paper 2011/08).

Ares, E. (2012). *EU ETS and aviation*. Retrieved from http://researchbriefings.files.parliament.uk/documents/SN05533/SN05533.pdf.

Brinke, L., & Faber, J. (2011). *The inclusion of aviation in the EU emissions trading system*. Retrieved from http://www.ictsd.org/downloads/2011/11/the-inclusion-of-aviation-in-the-eu-emissions-trading-system.pdf.

Calel, R., & Dechezlepretre, A. (2012). *Environmental policy and directed technological change: Evidence from the European carbon market*.

DEHSt. (2018). *Allocation of emission allowances to aircraft operators for trading periods 2012 and 2013–2020*.

Directorate-General for Climate Action. (2017). Retrieved from https://ec.europa.eu/clima/sites/clima/files/ets/auctioning/docs/auction_revenues_report_2017_en.pdf.

EUROCONTROL. (2015). *Annual report*. Brussels.

European Commission Climate Action. (2015). *The EU ETS handbook*. Retrieved from https://ec.europa.eu/clima/sites/clima/files/docs/ets_handbook_en.pdf.

European Commission Climate Action. (2018). *The EU Emissions Trading Scheme*. Retrieved from https://ec.europa.eu/clima/sites/clima/files/factsheet_ets_en.pdf.

European Energy Exchange. (2018). Retrieved from https://www.eex.com/en/.

European Parliament. (2018). Retrieved from http://www.europarl.europa.eu/news/en/headlines/society/20190313STO31218/co2-emissions-from-cars-facts-and-figures-infographics.

Heindl, P. (2012). *Financial intermediaries and emissions trading*.

IATA. (2015). *Airline disclosure guide*. Retrieved from https://www.iata.org/publications/Documents/Airline-Disclosure-Guide-aircraft-acquisition.pdf.

IATA. (2017). *Views on CORSIA and the application of EU ETS to aviation*. Retrieved from https://www.iata.org/policy/environment/Documents/iata-views-ets-corsia-feb17.pdf.

ICAO. (2016). *Resolution A39-3: Consolidated statement of continuing ICAO policies and practices related to environmental protection—Global Market-based Measure (MBM) scheme*. Retrieved from https://www.icao.int/environmental-protection/Documents/Resolution_A39_3.pdf.

United Nations Climate Change. (2018). Retrieved from http://unfccc.int/kyoto_protocol/mechanisms/clean_development_mechanism/items/2718.php.

CHAPTER 8

Impact of Air Transport on the Noise Level in the City of Sofia

Stela Todorova and Kaloyan Haralampiev

1 Introduction

Aviation is one of the fastest growing global industries. Today, the sector has grown remarkably, with passengers and cargo being carried no less than 49 billion kilometers annually, or 10 times the distance from Earth to Neptune. With over 1400 scheduled airlines, 26,000 aircraft in service, 3900 airports, and 173 air navigation services providers, aviation has established an unmatched global network at the service of travelers and businesses in every corner of the world. Report by the World Economic Forum (WEF) and Booz & Company estimates that the number of passengers traveling by air increased in the last two decades at a rate of about 5% per year (Chiesa and Ringbeck 2009).

It is also the safest and fastest means of transportation available, overcoming oceans and borders to connect people and support sustainable

S. Todorova (✉)
Department of Management and Marketing,
Agricultural University, Plovdiv, Bulgaria

K. Haralampiev
Department of Sociology of the Faculty of Philosophy,
Sofia University, Sofia, Bulgaria
e-mail: k_haralampiev@phls.uni-sofia.bg

T. Walker et al. (eds.), *Sustainable Aviation*,
https://doi.org/10.1007/978-3-030-28661-3_8

economic growth wherever aircraft fly. Current forecasts indicate that air traffic volumes will double in the next 15 years, characterized by a 4.6% annual growth rate for passenger traffic and 4.4% for freight traffic. The increased connectivity that it delivers leads to further reinvestment in aviation.

Since 1997, the EU air services market has been almost completely liberalized and harmonized. The new market organization allows every carrier to operate flights on any EU route. It also allows them to set their own tariffs, provided that the principles are fair competition.

The dynamics of flight connections and the reduction in ticket prices are the result of the liberalization of air transport regulations. The increase in demand for air services was a further consequence of these changes. Reduction of airfares, increased availability of air services, and rapid growth of network connections were the consequences of the development of the competition. Almost all airlines in Europe began to develop rapidly, in particular the low-cost carriers offering short and middle-distance flights (ICAO 2007).

Air transport is one of the major contributors to the growth in global gross domestic product. It also generates significant employment in all countries. Despite the global economic recession and the apparent crisis in air transport, its growth continues to be a global phenomenon with an average expected annual growth rate of 4.2–5.1% (ABB 2016, ATAG). Report of the International Civil Aviation Organization (ICAO) shows similar estimates. The same document reported estimations that by 2025, air passenger traffic in the world will increase by 4.6% per year while the total mileage in civil aviation will increase by 4.1% per year (ICAO Economic Development). The number of flights will increase three or even four times by 2035 if the growth rate of the number of air travelers continues to inflate (Chiesa and Ringbeck 2009).

Aviation, as every type of transport, is responsible for a number of negative external effects that are not neutral to the environment (Jakubiak 2015). Aviation is responsible for a small extent of a global environmental pollution, estimated approximately at 2–3%. This pollution is generated mainly in the area of airports (Jeż 2010).

The main goal of this paper is to reveal the relation between the noise level in the urban areas near Sofia airport and the air traffic. Our main research tasks are:

- To make a literature review of the problem;
- To gather data for the noise levels;
- To gather data for the air traffic;
- To choose relevant statistical methods and models for the revealing of the relation between the noise level and air traffic;
- To draw conclusions and to make recommendations about the noise pollution.

2 Literature Review

Aviation noise reduction is a problem that occurred in the middle of the last century in connection to the intensive operation of jet airplanes. It is still relevant and is an important part of the issues of ensuring human safety for the environment. In recent decades, the development and production of prospective civilian and military aircraft rely on environmental noise reduction requirements, which are regulated by the ICAO (Directive 2002/30/EC; NAREDBA № 16 from 14 January 1999).

Civilian and military aircraft produce significant amounts of sound, the majority of which is produced from either a turbofan or a turbojet engine (Manci et al. 1988; Visser and Wijnen 2001). Such engines generate relatively high amounts of vibration and turbulence. The main source of sound emanating from a jet aircraft results from the waves produced by the expanding gases of the jet exhaust stream (Manci et al. 1988). One of the most familiar and possibly most severe forms of aircraft noise is the sonic boom. This term describes the propagation of positive and negative air pressure generated by the aircraft traveling at speeds from one to five times the speed of sound in air (Pepper et al. 2003). Rotary-wing aircraft or helicopters are also a source of severe low-frequency sound and vibration (Delaney et al. 1999).

Environmental noise is an unwanted or harmful external sound caused by human activity, including the noise emitted by vehicles from road, rail, water, and air transport, from industrial installations and equipment as well as from local sources of noise. Noise pollution is generally defined as the exposure of people or animals to levels of sound that are annoying, stressful, or damaging to the ears. Humans are exposed to noise constantly, from barely audible levels to potentially painful and damaging levels.

3 Aircraft Noise and Humans

Depending on the subjective response to noise, people are divided into insensitive, sensitive, and hypersensitive. The first "suffers" the noise, the second expresses dissatisfaction, and the third suffers in great discomfort. It is also possible to switch from one group to another, depending on age, mood, health, occupation or holiday, etc. This subjective element is also reflected in the Environmental Noise Act and in the relevant regulation by introducing the terms "unwanted sound" and "discomfort."

However, objective negative impacts can be identified on people who are exposed to the impact of urban noise. The lighter effects of the noise are manifested by dull headache, dizziness with a sudden change in the situation, rapid fatigue, loss of concentration, sweating, drowsiness, or insomnia. With prolonged impact of the noise, more severe mental injuries (depressions) and permanent hearing damage occur.

Regardless of the momentary reaction, noise negatively impacts the ability and creativity of the modern human, especially when concentration and power of will are critical. For example, experts have found that weightlifters are worsening their situations due to the strong noise of throwing the bar on the podium by the other competitors in the training room.

The worst effect of noise is the partial or complete hearing loss (deafness). Deafness is mainly the result of a constant and excessive exposure to noise and an inability to process it in the inner ear, and more rarely caused by damage to the auditory nerve (nervous deafness).

According to the World Health Organization, about 5% of the Earth's population is hearing impaired. This percentage is higher in developed countries: between 6 and 8%. In these countries, background sound has increased in the cities by ¼ compared to the 1980s. It is also interesting to observe that in elderly people, deafness is the second most frequent symptom of arthritis.

The use of environmental influences is a vital response to the organism. Unlike eyesight, the hearing organ does not physiologically adapt to noise. However, there is a psychologically conditioned addiction to noise, for example, among residents of airports, motorways, some industrial plants, and so on.

It should be noted that there are sounds that people cannot get accustomed to (single strong blows, shots, blasts, etc.). Even though people do not feel noise, it can still be a health-damaging factor.

Observations show that the impact of sound is determined by different characteristics, such as the intensity of sound. The following negative effects have been identified at different noise levels:

- disturbances in mental work at 55 dB;
- responses of the vegetative nervous system at 65 dB;
- strong reactions of the vegetative nervous system at 80 dB;
- certain hearing damage (8-hour working day unprotected ear) at 85 dB;
- significant risk of hearing damage (unprotected ear, 8-hour working day) at 90 dB.

Aircraft noise is one of the most commonly reported residential disturbances. Noise annoyance has been defined as a feeling of resentment, displeasure, and discomfort when noise interferes with a person's thoughts, feelings, or actual activities (Passchier-Vermeer and Passchier 2000). Certain people are more susceptible to noise disturbance than others. Susceptibility may depend on quantitative factors such as the proximity or frequency of exposure to the sources of noise. Qualitative factors such as irritability, ethnic background, education level, and socioeconomic level may also affect susceptibility (Pepper et al. 2003).

Predicting the human response to noise is very difficult, and how humans perceive and respond to noise may depend on several factors, including intensity, frequency, and fluctuations in sound pressure. The given response is linked to emotional states, and humans will elicit a positive response when the noise generated is associated with a vital task. For example, imagine a person hearing a loud siren in the middle of the night. The siren noise is perceived tolerable (i.e., positively) because the person hearing the noise assumes that it is for the public good (ambulance, fire truck, or police car) (Pepper et al. 2003). Not surprisingly, long-term, uncontrolled exposure to aircraft noise may be perceived as negative. A negative perception is also dependent on physical variables, including the time of day, the type of neighborhood, season, and atmospheric conditions; cold, dense air conducts sound better than hot, breezy air. One will probably tolerate the airplane noise for reasonable periods of time, such as daylight hours, but that person is likely to complain if he or she is living in an area that is exposed to the constant flight paths of aircraft (Pepper et al. 2003). Aircraft noise may also be perceived negatively due to its intermittent, unpredictable nature (Suter 1994). This may be due to a process known as habituation, which is the term used to describe adaptive mechanisms to continuous stimuli (Boudreau 1968).

4 Studies Concerning Aircraft Noise

4.1 Human Studies

Generally, human effects commonly associated with noise pollution include speech interference (Chen and Charuk 2001), sleep deprivation, and hearing loss. Despite a lack of proof of causal relationships, research has attempted to link physical deformities to aircraft noise exposure (Jones and Tauscher 1978). Frequently exposed individuals seem to be easily distracted and unable to complete assigned tasks (Broadbent 1978). Other studies have attempted to correlate noise exposure to increased rates of hypertension (Rosenlund et al. 2001). These studies seem to support a report of increased blood pressure in rhesus monkeys (Muldoon and Miller 1989) exposed to chronic noise. In an earlier study, the negative impact of noise was evaluated using data collected from local doctors, and it suggested a causal relationship between self-reported health problems and noise exposure (Knipschild 1977).

A large number of studies emphasize negative health effects of air traffic on populations living in the vicinity of airports (Turnovska et al. 2004). They point out that long-term exposure to noise caused by aircraft disturbs everyday life and leads to increased irritability and impaired quality of life (Fidell et al. 2002; Haines et al. 2001; Kressin 1999; Smith and Holmsen 1998), reduction in working capacity, disturbance in the learning process, impairment of memory abilities (Hatfield et al. 2002; Hygge et al. 2002; Haines et al. 2001; Rosenlund et al. 2001), and serious difficulties in communication, especially when using a foreign language (Shimizu et al. 2002). Literature reports cardio-vascular disturbances (Rosenlund et al. 2001; Stansfeld et al. 2000; Turnovska et al. 2004) and the risk of mental disorders and other diseases (Rosenlund et al. 2001; Stansfeld et al. 2000; Hattori 2000). Loss of hearing has been reported in places with high levels of noise intensity, such as military airports (Miyakita et al. 2001). An extensive study of children attending schools near Heathrow Airport in London showed that the children were running a greater risk of cognitive impairment due to exposure to aircraft noise (Haines et al. 2001). The effect of noise on sick children is definitely unfavorable, even in short-term exposure. It was found that the noise of emergency helicopters transporting sick babies was a strong stress-producing factor (Buckland 2001). When takeoff and landing, airplanes are a source of infrasound (Nikolaeva and Paunov 1995) which, besides direct adverse health effect, may cause window panes to rattle, which in turn may cause irritation and anxiety among people exposed who live close to airfields (ICAO 2007). This latter effect is important, and house glazing is generally insufficient in reducing noise of this kind (Turnovska et al. 2004).

In their study about the Burgas Airport, Turnovska et al. (2004) reported that the diseases in populations living close to airports which, according to literature, might have an etiology in aircraft noise, are of paramount interest. Psychic disorders, psychoneuroses, and sense organs have been reported to be in positive dependence with the stress-producing influence of aircraft noise. The degree of involvement also correlates with the intensity and the duration of exposure.

5 Noise Effects on Wild Animals

Overall, many of the studies have evaluated the effects of noise on cardiac response (Weisenberger et al. 1996), biochemical responses (Jurtshuk 1959), and deep body temperature (Bakken et al. 1999) of various animal species. Some studies are mainly focused on the damage associated with exposure to aircraft noise on wildlife populations. For example, military activities have triggered redness in bald eagles (Haliaeetus leucocephalus) (Stalmaster and Kaiser 1999). Studies have also established the number of collisions between aircraft and birds (Burger 1983; Linnell et al. 1996).

Animals often rely on hearing to avoid predators, obtain food, and reproduce (Manci et al. 1988). For that reason, animals exposed to loud noise may exhibit changes in behavior patterns. In their studies, Boudreau (1968), Weisenberger et al. (1996), and Krausman et al. (1998) consider that animals may also adapt or habituate to noise exposures. Studies of the effects of noise exposure in animals have shown that "fear" is a common response (Bond et al. 1974). During a fear response, the animal tries to escape the noise by running or flying away. A factor associated with the initiation of a fear response in animals appears to be the distance of the aircraft from them. One study notes that an aircraft at a distance greater than 500 m from an animal does not cause any response (Ellis and Ellis 1991). Another study evaluating the response of mountain sheep to the noise of the aircraft shows an increase heart rate, but fails to conclude that it is harmful to mountain sheep (Krausman et al. 1998). These collective findings suggest that wild animals should be expected to adapt or avoid areas with a high noise impact.

6 Overview of Airports in Bulgaria

Aviation noise is typical of modern and medium-sized cities, near which there are civil airports. This noise is generated by the operation of the airports and is due to the aircraft, mainly on their takeoff and landing.

The aviation noise problem is complicated by the joint action of several factors. Most of the existing civil airports are refurbished and/or expanded airports that were built 40–50 years ago and sized for quite a few other aircraft and less traffic. Meanwhile, cities have grown and residential quarters have taken up part of the originally left buffer zone. Therefore, the classic solution—leaving a sanitary belt around airports—is practically difficult to apply.

Significant difficulties are also encountered in the design and construction of new airports. The distance from the settlements reduces the noise but increases the time and cost of travel from the airport to the serviced city. In some cases, time is commensurate with the duration of the flight. For the time being, major efforts have been made to reduce the level of aviation noise with a set of constructive, organizational, and spatial events.

The construction and maintenance of Bulgaria's transport infrastructure are based on the following strategic documents: Integrated Transportation Strategy 2030, Strategy for Development of the Transport Infrastructure of the Republic of Bulgaria until 2015, Operational Program "Transport," and General Plan for Transport (Master Plan). A well-developed airport infrastructure is of great importance for the development of tourism in the country. The country's geographic location predetermines it to be the point of departure and the destination, mostly of flights from and to Western Europe. For the period 2007–2017, a gradual increase in the number of passengers passing through the Sofia, Burgas, Varna and Plovdiv airports could be observed.

For the period between 2005 and 2014, the total passenger flow through airports in the country increased from 5035.9 thousand to 7795.4 thousand passengers, i.e., it has gone up by 1.55 times (Petkov et al. 2015, p. 73). In this context, the overall increase in the flow within the country stems from the increase in short-haul flights. According to Stoyanov's study, the increase in passenger flow of 116.13% in Bulgaria over the period between 2007 and 2012 is mainly due to the increase in short-haul flights (116.4%). The domination of short-haul transport in the Bulgarian aviation market means that 95.5 and 95.2% of flights, respectively, in 2012 and 2007, were with a duration of less than 3 flight hours (Stoyanov 2014). Therefore, only 5 out of 100 passenger flights had a destination further than 3 hours away. This confirms that the profile of the national consumer preferences for the usage of air transport is formed under the conditions defined by its convenience, speed of transportation, focus on local business, unorganized tourism, etc. Bulgaria's major airports are Sofia, Varna, Burgas, and Plovdiv. Our research focuses on the first airport.

Sofia Airport is the main international airport in the country with the largest passenger flow. Its 80th anniversary was celebrated in 2017. The airport has an asphalt-paved runway of 3600 × 45 m. The runway is fully equipped and has ICAO Category III B, which means that there are navigational aids installed on it that enable landing operations under low-visibility conditions, allowing the use of instrument climbing with a horizontal visibility of 150 meters and a vertical visibility of 15 meters. It has two passenger terminals, the second of which is built on an area of 56,500 sq. m. It is equipped with 7 Passenger Boarding Bridges and was commissioned at the end of 2006.

The planned annual traffic is 2.6 million passengers. It has a capacity of 2000 passengers. And it complies with modern standards for passenger (category "C" in IATA) and airline servicing. Terminal 1 was built in the first half of the twentieth century and has been extensively expanded and upgraded; it received an overhaul in 2000. Its total area is 9800 sq. m. Its declared operating capacity is 1.8 million passengers per annum.

The economic outcomes of the operation of Sofia airport provide significant revenue for Sofia Airport EAD (revenue from financing, ground handling, and other commercial activities), as well as for the Ministry of Transport (revenue from airport taxes). Sofia airport EAD provides job for more than 2000 employees. It also provides opportunities for tourism, business, and other trips to a significant number of citizens and foreigners. It also provides connection between Bulgaria and other countries.

The assessment of the health risk of the affected population can be seen in several aspects: risk during project implementation (during construction) and during its operation: direct risk, indirect risk, and total risk.

During the construction period, the population of the closest sites subject to health protection is mainly exposed to dust (excavation and bulk operations), noise, exhaust gases from construction and transport equipment, welding aerosols, and more. If the design of the airport does not specify in particular the type and quantity of the construction works, an estimate of this risk cannot be made. However, experience can be predicted that at particular moments of adverse weather conditions (wind from the east or west) it is possible to exceed the regulated air pollution in the nearest objects subject to health protection. Overall, however, the risk factors during the construction period mainly affect the territory of the airport complex and no significant adverse health effects of the potentially affected population are expected. The main and most significant adverse effects when looking at the health risks of the population are related to the efficacy of the operation of Sofia, Varna, and Burgas Airports as a whole. The importance of noise is a key factor in the health and the functions of the affected person, as well as the risks to health.

7 Data and Methods

Our data sources for the noise level are Quarterly Reports with Results of Continuous Measurements of the Noise Giving a Notion for the Acoustic State of Areas Located near Sofia Airport.[1] The data is on monthly basis and covers the period from October 2015 to December 2017, i.e., 27 months. We received data for August and September 2015 directly from Sofia Airport. However, this data is on a daily basis. Therefore, we calculated the monthly average value using the following equation:

$$\bar{x} = 10.\log_{10}\left(\frac{\sum_{i=1}^{N} 10^{\frac{x}{10}}}{N}\right)$$

[1] http://eea.government.bg/bg/dokladi/threemonth/.

Fig. 1 Noise registration location points

where x are the daily values and N is the number of the days.[2]

In the period April–July 2015, the noise monitoring system was updated. Thus, we have data only for 29 months.

Therefore, we gathered data for the air transport for these 29 months. The data source is statistical information about the International Airports in the Republic of Bulgaria.[3]

The noise indicators are:

- Twenty-four-hour average equivalent level of cumulative noise;
- Twenty-four-hour average equivalent level of noise due to flights;
- Twenty-four-hour average equivalent level of background noise.

The noise is registered in eight location points—six west of the airport and two east. They are denoted as W1, W2, W3, W4, W5, W6, E1, and E2. Larger numbers show a larger distance from the airport. However, there are full-time series only for W1, W2, W3, and W4. In addition, there are only two missing values for E1, and we interpolated these two missing values by exponential smoothing. For the other three location

[2] http://plovdiv.riosv.com/files/file/noise/Naredba6.pdf.

[3] https://www.caa.bg/bg/category/292/statisticheska-informaciya.

points, there are more missing values and because of that we excluded these location points from our data set. As a result, we have data for five location points that are presented in Fig. 1.

In our analysis, we use only the first two noise indicators as dependent variables, and we take the average mean of all five location points each month. Thus, we measure the noise that affects approximately 145,000 people or approximately 11% of the Sofia population.[4]

The air traffic indicators are:

- Aircraft movements (thousands)
- Passengers (thousands)
- Freight (tons)
- Mail (tons).

We use all these four traffic indicators as independent variables.

Our data set consists of time series. However, "the correlation between such series could prove deceivably strong if the trends in these series are identical by chance" (Velichkova 1981; citing Yule and Kendall 1950). In such case, we need to get rid of the so-called autocorrelation, i.e., the influence of the former state of the researched phenomenon on its present state. There are many ways for doing so, but the most elegant and easiest way is to introduce time in the model. Introducing time in the model is based on a corollary of the Frisch-Waugh theorem, which states that the "correlation of the residuals around trend lines is equal to the establishment of a multi-factor regression model for the expression of the dependence between the absolute values of the Y and X series with the additional factor of time t" (Velichkova 1981).

As long as the regression model is built upon time series, we introduce "the time" as an additional factor in our model, which provides concrete practical advantages: (1) simplifying the calculation procedures and (2) providing the opportunity to account for all other factors that we do not control (Velichkova 1981).

[4] According to the National Census conducted in 2011.

8 Results

8.1 Noise Pollution

The noise levels in the urban areas near Sofia airport are presented in Fig. 2.

According to the chapter noise pollution in the National Report on the State and Protection of the Environment in Bulgaria,[5] noise levels of 50–80 dB make it difficult to hear speech, and noise levels of about 50–60 dB have a harmful effect on the people's nervous system and disturb their work and rest. Unfortunately, Fig. 2 shows that the noise levels in the urban areas near Sofia Airport are above 50 dB and that even the cumulative noise levels are above 65 dB. It is thus important to study the factors that affect the noise level in order to find solutions for its reduction.

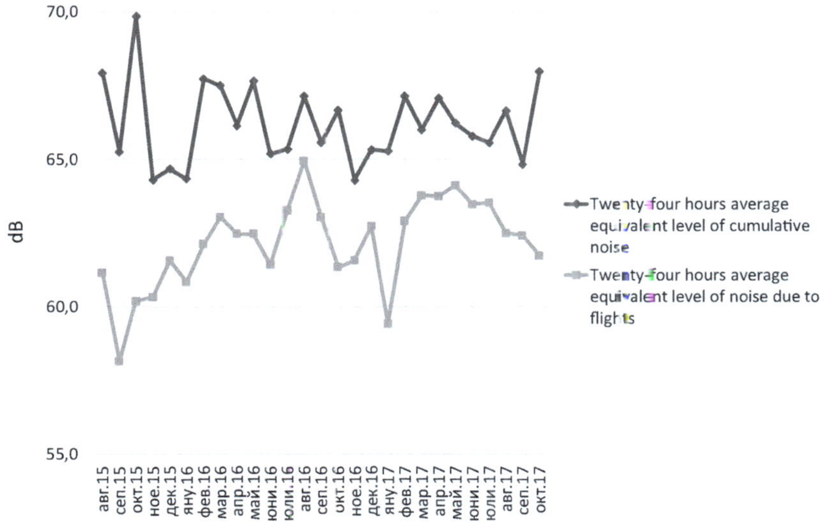

Fig. 2 Noise levels in the urban areas near Sofia airport

[5] http://eea.government.bg/bg/soer/2014/noise/index.

8.2 Regression Model with Dependent Variable Twenty-Four-Hour Average Equivalent Level of Cumulative Noise

Collinearity diagnostics show that the independent variable passengers (Thousands) depends on all others independent variables and must be excluded. The new model is adequate ($F=3.14$, $p=0.033$), and its multiple R^2 is 0.344, i.e., 34.4% of the variance of the twenty-four-hour average equivalent level of cumulative noise is explained by all independent variables together. The coefficients of the new model are presented in Table 1.

Increasing the aircraft movements by 1000 results in an increase in the twenty-four-hour average equivalent level of cumulative noise by 1.88 dB, provided that the other factors in the model do not change.

Increasing the mail by one ton results in an increase in the twenty-four-hour average equivalent level of cumulative noise by 0.03 dB, provided that the other factors in the model do not change.

The twenty-four-hour average equivalent level of cumulative noise decreases by 0.12 dB monthly, provided that the other factors in the model do not change.

Freight does not affect statistically significant the twenty-four-hour average equivalent level of cumulative noise.

8.3 Regression Model with Dependent Variable Twenty-Four-Hour Average Equivalent Level of Noise Due to Flights

Collinearity diagnostics show that the independent variable passengers (Thousands) depends on all others independent variables and must be excluded. The new model is adequate ($F=7.06$, $p=0.001$), and its

Table 1 Coefficients of the regression model with dependent variable twenty-four-hour average equivalent level of cumulative noise

Independent variables	*Unstandardized coefficients*	*Significance level*	*Partial correlations*
Aircraft movements (thousands)	1.880	0.020	0.454
Freight (tons)	0.001	0.207	0.256
Mail (tons)	0.034	0.016	0.467
Time	−0.120	0.018	−0.462

Table 2 Coefficients of the regression model with dependent variable twenty-four-hour average equivalent level of noise due to flights

Independent variables	*Unstandardized coefficients*	*Significance level*	*Partial correlations*
Aircraft movements (thousands)	1.685	0.025	0.440
Freight (tons)	0.001	0.400	0.172
Mail (tons)	−0.012	0.332	−0.198
Time	−0.006	0.900	−0.026

multiple R^2 is 0.541, i.e., 54.1% of the variance of the twenty-four-hour average equivalent level of noise due to flights is explained by all independent variables together. The coefficients of the new model are presented in Table 2.

Increasing the aircraft movements by 1000 results in an increase in the twenty-four-hour average equivalent level of noise due to flights by 1.69 dB, provided that the other factors in the model do not change.

All other independent variables do not statistically and significantly affect the twenty-four-hour average equivalent level of noise due to flights.

9 Discussion and Mechanisms to Reduce Aircraft Noise

Only one independent variable (aircraft movements) affects the twenty-four-hour average equivalent level of noise due to flights. It is easy to explain since the noise during takeoffs and landings is almost the same, regardless of the amount of cargo and mail carried by the airplanes. Rather, it depends on the types of aircraft.

Concerning the twenty-four-hour average equivalent level of cumulative noise, it is easy to explain why the freights and mail appear as statistically significant factors together with aircraft movements, and that is because the transportation, loading, and unloading of the freights and mail produce additional background noise.

Since the aircraft movements are the most important factor—and we expect their increase in the future—this would lead also to the increasing of the noise levels. For that reason, other technical solutions for reducing noise levels should be sought.

Aviation noise is more irritating than noise from other sources for two reasons. The usual noises in the residential environment are relatively constant (ground transport, noise from household appliances, etc.), while aircraft noise consists of sporadic noise events with a certain rising and falling pattern. People do not perceive it as yet another component of the daily "background noise" and every flyover is noticeable and irritating. Another characteristic that determines this impact of aircraft noise is its higher frequency and power to which the human ear is particularly sensitive (Lazarova 2005).

Reducing the level of aircraft noise is an environmental issue central to the activities of international environmental and aviation organizations. The reduction of noise from airports should be sought in two ways: (1) a good urban development solution that would keep the city away from the airport. For Sofia Airport, this requirement is to a certain extent respected and (2) making decisions regarding the design of the airplane, the engine, and the path of its movement.

The solutions for reducing aircraft noise are:

1. Selecting new "noisy" engines, reducing night flights, investing in noise abatement technologies, applying fines, and operating restrictions and quotas. At night, the quietest aircrafts are allowed to land and fly, and they are further limited by a point system known as a quota reporting system. For each landing and take-off, airplane noise level points accumulate, with cumulative points not exceeding a certain amount over a certain period of time. Many airports charge landing and take-off fees based on the results of aircraft noise certification (Tochev et al. 2014).
2. Reduce the resistivity of the aircraft by reducing the required traction. Improving the aerodynamic efficiency of the aircraft by reducing resistance or thrust respectively reduces the amount of fuel required and emissions (Tochev et al. 2014).

In connection with the increase in air traffic at Sofia Airport, BULATSA, the Civil Aviation Administration, and Sofia Airport EAD took additional measures to limit air noise to Sofia Airport, which entered into force on 26 April 2018. The measures taken are the results of a targeted policy of the three institutions to reduce the impact of aviation on the environment in the conditions of increasing traffic. The main priority is the safe operation of the flights in the interest of the passengers, the residents of Sofia, and the surrounding settlements.

A procedure is introduced to reduce aviation noise in the eastern direction, which enables for the airplanes' takeoffs heading to Western European airports to raise an absolute altitude of 4500 feet/1370 meters and then turn left. The measure does not allow to fly over the village of Dolni Bogrov, and if some of them pass, they will be at a height of 1000 meters above the buildings, instead of 400 meters as it has been so far.

Aviation noise is reduced when the aircraft is stopped after landing at the Sofia Airport runway. The procedure takes into account the increased runway length and enables the aircraft to keep going after landing using the maximum reverse engine thrust. This allows noise reduction over the neighborhoods near the airport (Hristo Botev, Suhata Reka, Vasil Levski).

A procedure to prohibit deviations from established flight paths to reduce the length of the route at an absolute height of less than 5000 feet/1500 meters will restrict the practice immediately after takeoff to take the airplanes to the shortest route to landing. The measure will reduce aircraft noise over Sofia and Dolni Bogrov.

10 Conclusion

Our study shows that noise levels in urban areas near Sofia Airport are above 50 dB, and even cumulative noise levels are above 65 dB. The aircraft noise affects 145,000 people or approximately 11% of the Sofia population.

National legislation concerning aircraft noise is oriented toward the introduction of standards for aircraft noise, detecting and mapping the distribution of noise around airports, collecting data for aviation through closed (not publicly oriented) monitoring systems, and distributing non-operational information for the observed noise (Petkov 2017).

Consequently, operating aircraft at Sofia Airport generate noise above the norm in the districts "Hristo Botev," "Suhata Reka," "Nova mahala," and more. The new runway at Sofia Airport, with its departure from the residential areas next to it and the change of aircraft with those with lower-noise level, reduces the noise by up to 40% compared to its condition before the reconstruction and expansion. It is necessary to control flight management in compliance with the meteorological conditions and the clearing of the adjacent territories of Sofia Airport based on the data from the monitoring system, located along the take-off/landing route.

Noise caused by human activity, including aircraft noise, does not directly lead to pollution of the natural environment, but directly impacts the quality of life.

References

ABB Annual Report. (2016). ATAG.

Bakken, M., Moe, R. O., Smith, A. J., & Selle, G. M. E. (1999). Effects of environmental stressors on deep body temperature and activity levels in silver fox vixens (vulpes wipes). *Applied Animal Behaviour Science, 64*(2), 141–151.

Bond, J., Rumsey, T. S., Menear, J. R., Colbert, L. I., Kern, D., & Weinland, B. T. (1974). *Effects of simulated sonic booms on eating patterns, feed intake and behavioral activity of ponies and beef cattle.* International Livestock Symposium, Department of Agriculture, ARS, BARC-East, Beltsville, Maryland, SP-0174, American Society of Agriculture Engineers, USA.

Boudreau, G. W. (1968). Alarm sounds and responses of birds and their application in controlling problem species. *Living Bird, 7,* 27–46.

Broadbent, D. E. (1978). The current state of noise research: Reply to Poulton. *Psychological Bulletin, 85,* 1052–1067.

Buckland, E. (2001). Transport noise stresses sick babies. *Australian Nursing Journal, 9,* 41.

Burger, J. (1983). Jet aircraft noise and bird strikes: Why more birds are being hit. *Environmental Pollution Series A Ecological and Biological, 30*(2),143–152.

Chen, A. C., & Charuk, K. (2010). Speech interference levels at airport noise impacted schools. *Sound and Vibration, 35*(7), 26–31.

Chiesa, T., & Ringbeck, J. (2009). *Towards a low carbon travel & tourism sector.* World Economic Forum Report prepared with the support of Booz & Company.

Delaney, D. K., Grubb, T. G., Beier, P., Pater, L. L., & Reiser, M. H. (1999). Effects of helicopter noise on Mexican spotted owls. *Journal of Wildlife Management, 63,* 60–75.

Direktiva 2002/30/EO na Evropeiskia parlament i na Syveta na evropeiskia syjuz otnosno ustanoviavaneto na pravila i procedure za vyvezhdaneto na eksploatatsionni ogranichenia, svyrzani s shuma na letishtata na Obshtnostta.

Ellis, D. H., & Ellis, C. H. (1991). Raptor responses to low-level jet aircraft and sonic booms. *Environmental Pollution, 74,* 53–83.

Fidell, S., Silvati, L., & Haboly, E. (2002). Social survey of community response to a step change in aircraft noise exposure. *Journal of the Acoustical Society of America, 111*(1), 200–209.

Haines, M. M., Stansfeld, S. A., Brentnall, S., Head, J., Berry, B., Jiggins, M., et al. (2001). The West London Schools Study: The effects of chronic aircraft noise exposure on child health. *Psychological Medicine, 31*, 1385–1396.

Hatfield, J., Job, R. F., Hede, A. J., Carter, N. L., Peploe, P., Taylor, R., et al. (2002). Human response to environmental noise: The role of perceived control. *International Journal of Behavioral Medicine, 9*(4), 341–359.

Hattori, M. (2000). A field study of health effects of aircraft noise in adults around Komatsu Air Base (1998). *Nippon Koshu Eisei Zasshi, 47*(1), 20–31.

Hygge, S., Evans, G. W., & Bullinger, M. (2002). A prospective study of some effects of aircraft noise on cognitive performance in schoolchildren. *Psychological Science, 13*(5), 469–474.

ICAO Environmental Report. (2007).

Jakubiak, M. (2015). Environmental impact of air transport—Case study of Krakow Airport. *Logistyka, 2*, 276–283.

Jeż, M. (2010). *Ekologiczna problemy portu lotniczego* (Prace Instytutu Lotnictwa Nr 4 [206]). Wydawnictwa Naukowe Instytutu Lotnictwa, pp. 59–71.

Jones, F. N., & Tauscher, J. (1978). Residence under an airport landing pattern as a factor in teratism. *Archives of Environmental Health, 33*(1), 10–12.

Jurtshuk, P. (1959, August). Endocrine changes due to auditory stress. *Acta endocrinologica, 31*(3), 405–418.

Knipschild, P., VI. (1977). Medical effects of aircraft noise: Community cardiovascular survey. *International Archives of Occupational Environmental Health, 40*(3), 185–190.

Krausman, P. R., Wallace, M. C., Hayes, C. L., & De Young, D. W. (1998). Effects of jet aircraft on mountain sheep. *Journal of Wildlife Management, 62*, 1246–1254.

Kressin, J. (1999). Aviation medicine problems in otorhinolaryngology. *Zeitschrift fur arztliche Fortbildung und Qualitatssicherung, 93*(7), 509–512.

Lazarova, A. (2005). *Aviation noise*. Sofia: New Bulgarian University.

Linnell, M. A., Conover, M. R., & Ohashi, T. J. (1996). Analysis of bird strikes at a tropical airport. *Journal of Wildlife Management, 60*, 935–945.

Manci, K. M., Gladwin, D. N., & Villella, R. (1988). *Effects of aircraft noise and sonic booms on domestic animals and wildlife: A literature synthesis*. NERC 88/29, AFESC RE 88.14.

Miyakita, T., Yoza, T., Matsui, T., Ito, A., Hiramatsu, K., Osada, Y., et al. (2001). An epidemiological study regarding the hearing acuity of residents in the area with high level of aircraft noise: Results of hearing tests conducted in the vicinity of Kadena Air Base. *Nippon Eiseigaku Zasshi, 56*(3), 577–587.

Muldoon, J. P., & Miller, R. L. (1989). Low levels of aircraft noise from expanded east coast plan operations. *Inter-Noise* 89, 665–670.

NAREDBA № 16 ot 14 January 1999 za aviacionnia shum i za gazovite emisii na aviacionnite dvigateli. Izdadena ot ministyra na transporta.

Nikolaeva, D., & Paunov, I. (1995). Infrazvuk, ultrazvuk, shum i vibrasii. Infrasound. In D. Tzvetkov & M. Angelova (Eds.), *Infrasound, ultrasound, noise and vibrations* [in Bulgarian] (pp. 11–43). Sofia: Medicina i Fizkultura Publisher.

Passchier-Vermeer, W., & Passchier, W. F. (2000). Noise exposure and public health. *Environmental Health Respiratory, 108*, 123–131.

Pepper, C., Nascarella, M. A., & Kendall, R. J. (2003). A review of the effects of aircraft noise on wildlife and humans, current control mechanisms, and the need for further study. *Environmental Management, 32*(4), 418–432.

Petkov, T. (2017). Develop a draft legally binding model for implementation of sanctions/penalties when managing the impact of noise on living around airports. *Ecological Engineering and Environment Protection*, No. 3.

Petkov, T., Stanulov, S., & Yotsev, Y. (2015). Current review of Bulgarian aviation industry. *Journal of Mechanics, Transport and Communications, 13*(3/1), 71–73. http://www.mtc-aj.com.

Report of Clean Air Forum 16 and 17 November 2017. www.eea.government.bg/en/news/clean-air-forum-report.pdf.

Rosenlund, M., Berglind, N., Pershagen, G., Jarup, L., & Bluhm, G. (2001). Increased prevalence of hypertension in a population exposed to aircraft noise. *Occupational Environmental Medicine, 58*(12): 769–773.

Shimizu, T., Makishima, K., Yoshida, M., & Yamagishi, H. (2002). Effect of background noise on perception of English speech for Japanese listeners. *Auris Nasus Larynx, 29*(2), 121–125.

Smith, A., & Holmsen, E. (1998). Public health work in connection with a new international airport. *Tidsskrift for den Norske laegeforening: tidsskrift for praktisk medicin, ny raekke, 118*(23), 3656–3658.

Stalmaster, M. V., & Kaiser, J. L. (1999). Flushing responses of wintering bald eagles to military activity. *Journal of Wildlife Management, 61*, 1307–1313.

Stansfeld, S., Haines, M., & Brown, B. (2000). Noise and health in the urban environment. *Reviews on Environmental Health, 15*(1–2), 43–82.

Stoyanov, M. (2014, October). *Challenges of low-cost air carriers*. Scientific conference "Transport in a changing world—Challenges and solutions", Sofia, pp. 226–236.

Suter, A. H. (1994). *Report on the proposed reduction in noise levels at the Seattle-Tacoma Airport*. www.rcaanews.org/suter94.htm. Last visited 26 July 2002.

Tochev, K. et al. (2014). Construction solutions of reducing aircraft noise. *Mechanics Transport Communications, 3*(1). http://www.mtc-aj.com.

Turnovska, T., Staykova, J., & Petkov, T. (2004). Health assessment in populations exposed to aircraft noise. *Arhiv za higijenu rada i toksikologiju, 55*(1), 5–10.

Velichkova, N. (1981). *Statisticheski Metodi za Izychawane i Prognozirane Razwitieto na Sotsialno-ikonomicheski Jawlenija* [Statistical methods for studying and forecasting the development of socio-economic phenomena]. Sofia: Science and Art.

Visser, H. G., & Wijnen, R. A. A. (2001). Optimization of noise abetment departure trajectories. *Journal of Aircraft, 38*(4), 620–627.

Weisenberger, M. E., Krausman, P. R., Wallcace, M. C., De Young, D. W., & Maughan, O. E. (1996). Effects of simulated jet aircraft noise on heart rate and behavior of desert ungulates. *Journal of Wildlife Management, 60*, 52–61.

Yule, G. U., & Kendall, M. G. (1950). *An introduction to the theory of statistics* (3rd ed., pp. 162–163). Harper Publishing Company.

CHAPTER 9

Social Dimensions of Aviation on Sustainable Development

Dimitrios J. Dimitriou and Maria F. Sartzetaki

1 Introduction

Socio-economic development is defined as a multidimensional topic affecting many aspects of people's lives, directly linked with the sustainable development of business. Evidence on this is the international organizations, which provide dedicated activities to promote the goals of human development and well-being towards sustainable business development (ICAO 2017; IATA 2017; UNECE 2015; UNDP 2014; OECD 2013; World Bank 2013). World bank report (2013, pp. 3, 4) highlights the need to address gender inequality as a top goal (ranged 3rd) upon global socio-economic development improvement.

Sustainable development, in a broader sense, refers to an increase in a country's well-being. Economic development is based on both

D. J. Dimitriou (✉) · M. F. Sartzetaki
Department of Economics, Democritus University of Thrace (DUTh), Komotini, Greece
e-mail: ddimitri@econ.duth.gr

M. F. Sartzetaki
e-mail: msartze@econ.duth.gr

T. Walker et al. (eds.), *Sustainable Aviation*,
https://doi.org/10.1007/978-3-030-28661-3_9

economic and non-economic factors for the general improvement of living conditions and, in general, the raising of the level of prosperity. Economic growth is separate from the quality of life, without the first necessarily involving the latter (Léle 1991). Until nowadays, the most important indicator of economic growth is the long-term growth rate of GDP per capita, but due to its disadvantages, there is a necessity to introduce other indicators such as health, education, and longevity indicators. Analysing macroeconomic adjustment from institutional perspective demands a detailed explanation of how institutional options are and should be made. Highly valuable to this effort is the analysis of norms, values, and principles relating to economic agents, economic structures, and decision-making procedures (Hazakis 2017). The need for wider measures of sustainable development has been supported by key stakeholders in the global development arena that have demanded for a multidimensional concept of human well-being (ICAO 2017; IATA 2017; UNECE 2015; UNDP 2014; OECD 2011, 2013; World Bank 2013).

The proposed chapter deals with the challenges in sustainable development in aviation in terms of diverse social benefits. A key objective is to illustrate the aviation social benefits towards a sustainable development and the respectfulness of human values, including indicators such as human equality and employee equal opportunities. Key outputs are highlighted by a top-down approach, definitions, and assessment methodology. Conventional wisdom is to stimulate the interest on topic and promote key actions towards the social dimension of aviation in sustainable development.

2 Sustainable Development Dimensions

Sustainable development, a key concept in the twenty-first century, is a primary concern for decision-makers dealing with transport infrastructure development. Sustainable development is the common starting point for interaction between all “stakeholders”: national and local governments, authorities, and public and private sectors, such as non-governmental organizations and citizens. The principles of sustainable development were formally formulated for the first time at the 1992 Rio International Conference, which resulted in Agenda 21 (UN 1992, 1993). Sustainable development was initially defined as “development that provides long-term economic, social, and environmental benefits by taking care of the needs of present and future generations". The 1992 Maastricht agreement (EC 1992), the treaty of Amsterdam in 1997 (EC 1997) and the Johannesburg International Conference in 2002 (UN 2002), confirmed the necessity of sustainability and the systemic framework, incorporating sustainable development into European Union Law.

Despite the recent developments, there are still many trends and issues in several different areas that are not based on the principles of sustainable development. Demand for natural resources has increased rapidly, while biodiversity has declined overall, and larger ecosystems are receiving more and more pressure. Energy consumption in transport continues to rise, poverty persists, and therefore, more efforts are needed globally to achieve Millennium Development Goals set by UNECE (UN 2006). Moreover, the recent economic and financial crisis has shown that sustainability is a crucial factor for financial stability and economic development.

Based on comprehensive literature review, sustainable development is divided in three dimensions: the environment, the economy, and society. Each dimension is linked to the other two (UN 2007). Several different depictions of the concept of sustainable development have been proposed in the literature. A widespread depiction that represents sustainable development is through overlapping dimensions (Connelly 2007), (Fig. 1). Within the limits of the overlap of the three dimensions,

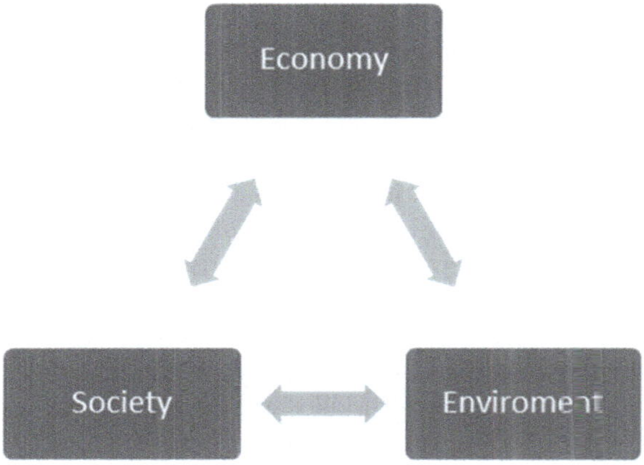

Fig. 1 Sustainable development dimensions depiction

the concept of sustainable development is indifferent to its three dimensions, thus reinforcing the view that sustainable development is a single goal interacting with the three dimensions (Jacobs 1999).

3 Review of Key Sustainable Development Goals Subject to Aviation

Globally, a series of major international developments have promoted the role of sustainable development (UNDP 2014). The Paris Agreement on Climate Change adopted by United Nations Framework Convention on Climate Change (December 2015) supports a pathway for transforming the world's mobility in the coming decades. This Paris Agreement (2015) has created the circumstances for the transport sector to develop a roadmap towards carbon neutrality.

Another very important development towards the path of transport and sustainable development is the adoption of the Agenda for Sustainable Development by the United Nations in 2016, which highlights the role of transport in economic development (ITF 2017). The 2030 Agenda is composed of 17 Sustainable Development Goals (SDGs) which are supported by 169 targets. All countries and all stakeholders, acting in collaborative partnership, will implement this plan. The 17 SDGs and 169 targets demonstrate the scale and ambition of this new universal Agenda for the sustainable development for 2030. The sustainable aviation is implicit, mainly in the goals SDG1, SDG2, SDG3, SDG4, SDG9, SDG10, and SDG13.

The sustainable development goals SDG 1: No Poverty, SDG 2: Zero Hunger and SDG 8: Decent Work and Economic Growth aim to promote sustained and inclusive economic growth, full and productive employment. Especially as refers to aviation, economic development is driven through improved connectivity, as it lifts standards of living, by creating jobs in the aviation industry, as well as other sectors indirectly, and especially supporting employment in tourism industry.

The SDG 3: Good Health and Well-being, aiming to ensure healthy lives and promote well-being subject to aviation, refers to safety and security, access to vital medical care in remote communities and transporting time-sensitive medical supplies and major role in disaster relief. The SDG 4: Quality Education, aiming to ensure inclusive and quality education for all, and promote lifelong learning subject to aviation, refers to the promotion of access to educational opportunities and quality education for aviation sector employees (UNECE 2015)

The SDG 9: Industry, Innovation, and Infrastructure, aiming to build resilient infrastructure, promote inclusive and sustainable industrialization, and foster innovation subject to aviation, refers to the supply of research and development expenditure as a proportion of per million inhabitants and the proportion of medium and high-tech industry value-added. Infrastructure investment covers spending on new transport construction and the improvement of the existing network. Infrastructure investment is a key determinant of performance in the transport sector. Inland infrastructure includes road, rail, inland waterways, maritime ports, and airports and takes account of all sources of financing. Efficient transport infrastructure provides economic and social benefits to both advanced and emerging economies by improving market accessibility and productivity, ensuring balanced regional economic development, creating employment, promoting labour mobility, and connecting communities. The SDG 10: Reduced Inequalities within and among countries subject to aviation refers to the reduction of inequality between countries, and between individuals, by creating trade links caused by improved connectivity (UNECE 2015).

The SDG 13: Climate Action to combat climate change and its impacts subject to the aviation sector refers to the achievement of sustainable development through the development of new technology, commercialization of sustainable aviation fuels, and more efficient navigational infrastructure. All sectors of the aviation industry have agreed on a robust strategy for reducing CO_2 emissions and are making excellent progress working towards global climate goals, such as the Carbon Offsetting and Reduction Scheme for International Aviation (CORSIA), successfully agreed in October 2016 by the UN International Civil Aviation Organization (ICAO) on the which will enable the industry's shared goal of carbon-neutral growth.

4 Demand and Supply Variables for Aviation Sustainable Development

Sustainable transport infrastructure is defined as a development process that incorporates the basic principles of sustainable development. These processes should comply with the objectives of environmental, social awareness, and economic growth (Dimitriou et al. 2017; Samberg et al. 2012). Institutions, associations, and governmental bodies widely recognize the need for monitoring transport demand and adopting strategies to exploit the total benefits for local society. Dimitriou and Sartzetaki (2018a) estimate the contribution of aviation in regional development providing evidence that this relationship needs to be investigated not only to extrapolate the demand trends, but also to adopt policies, define strategies, and support decisions towards transport policies and investments in new infrastructure to accommodate additional demand.

Transport is a major contributor to economic growth that requires operational productive and efficient infrastructures and services. At the same time, CC is likely to have essential implication for transport industry infrastructure. The need and the budget for the inclusion of adaptation measures in the transport industry are highlighted by UNECE (2013). Dimitriou et al. (2014) review the CC adaptation strategies and environmental mitigation actions for 15 European airports, highlighting the low performance of Greece and South Europe. Taking all these factors into account, a mitigation/adaptation chart (Fig. 2) is constructed to demonstrate that the supply/demand elements are two sides of the same coin. This is because mitigation and adaptation will impact the supply/demand relationship differently in the short and long run. Figure 3 divides the variables between classical (those which are used in existing

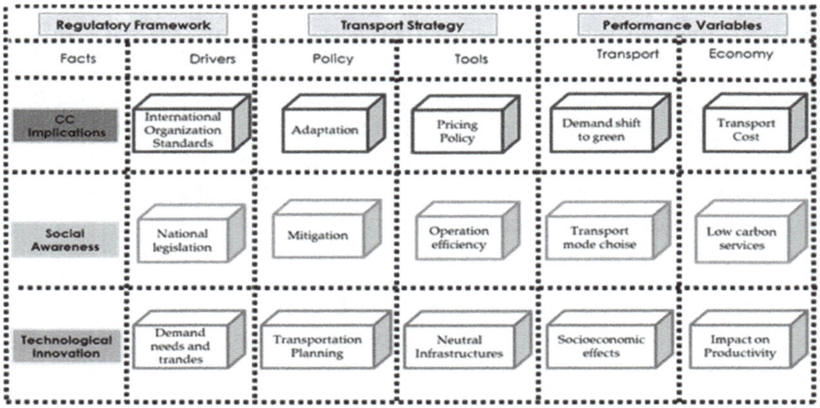

Fig. 2 Demand-supply variables influence the transport-regional economy equilibrium (Dimitriou and Sartzetaki 2016)

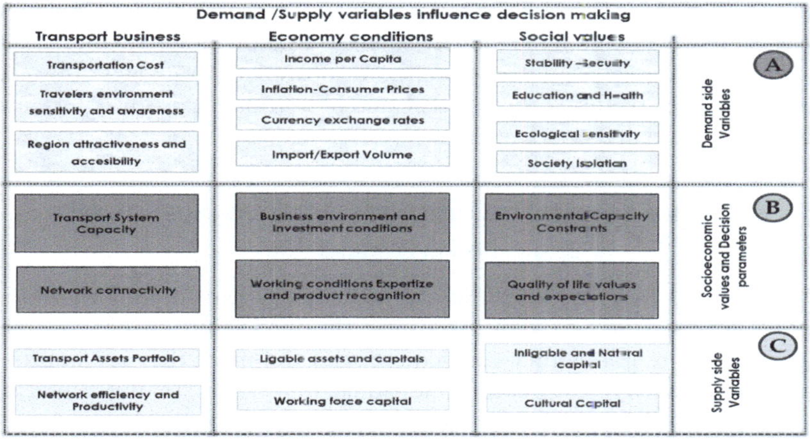

Fig. 3 Demand-supply variables influence the transport development decision-making (Dimitriou and Sartzetaki 2016)

tourism demand models in European destinations, and the new climate-related variables, which need to be included in the models.

The related demand variables, such as the transport cost, income per capita, social security, education, health and the related supply variables as transport asset portfolio, working force capital and other capitals, affect the decision-making process in order to develop the transport sector and lead to the equilibrium point between transport sustainability and economic development (Dimitriou 2016). Economic geography, a branch of economics, indicates that the values assigned to classical variables for the models that forecast tourism demand are unique to each country as illustrated in Fig. 2. These variables are self-explanatory and interrelated. For example, although natural and climatic capital may exist in many countries, it is a combination of the existing tourism infrastructure and historical levels of tourism that together with the natural and climatic capital explain why volumes of tourism are higher in certain countries equilibrium (Dimitriou and Sartzetaki 2017).

Indeed, the costs of trade, like trade in general, explain these patterns. A country's comparative advantage in market prices is a function of the costs associated with distance and infrastructure, and with the agency and transaction costs associated with industry and institutional structures. These accumulated comparative advantages are more complex than simply ambient temperature or even aviation infrastructure, they take time to develop and therefore the comparative advantages also take time to erode. In short, the classical variables assume existing assets and cannot be viewed as operating in a vacuum.

5 The Social Dimension of Aviation Sustainable Development

When evaluating the contribution of air transport, income is the main economic measure of living standards; the higher the income people have, the more goods they can acquire, and the more they are improving their financial position and individual well-being. In this sense, as according to Nobelist Robert Lucas (1988) "From the moment someone begins to think about the phenomenon of economic growth, it is very difficult to think of anything else" (Lucas 1988). The need for reorienting factors that affect economic development except that GDP per capita, by considering the development of the social dimension, so as not to limit the future growth potential has led to the concept of sustainable development. The development is much broader than economic growth and goes beyond economic, social, and cultural development and the environment. The definition of development includes not only the objectives, but also the means by which these objectives will be achieved (Sen 2000).

It is analysed in literature that the measurement of economic development should include many indicators. The so-called social indicators were developed in the 1960s, while later "environmental indicators" were included systematic efforts were made in the last decade to develop commonly accepted "sustainable development indicators" at European level. On 20 August 2009, the Commission published a significant commission communication with title "GDP and beyond: Measuring progress in a changing world", as many stakeholders recognize the need to supplement GDP with environmental and social indicators. This concept has been developed by United Nations Committees (Sen 2000).

To address these problems, a systematic effort is being made by economic science and by the field of social and environmental economics. The main directions of this effort can be characterized as (a) the internalization of environmental externalities, such as pollution, (b) the reassessment of the value of environmental services, which are systematically undervalued, and (c) problems stemming from the fact that many natural resources are available. Very crucial to this analysis is the literature review on the "greening" of National Accounts (Dasgupta et al. 2000).

Aviation is considered a major economic sector—one important and which in addition provides vital connecting infrastructure for the lives and livelihoods of a large share of the world's population. Aviation links markets and individuals, making regions more competitive and promoting individuals' social and economic development, and thus affecting the three pillars of sustainability: social, economic, and environmental (Dimitriou and Sartzetaki 2018b).

According to IATA (2017), many factors and forces ("drivers of change") must be considered when making decisions that shape the future of the aviation industry and, directly or indirectly, the lives and well-being of the millions of people who depend on it. International visions for aviation improvements need to fulfil wider objectives, including not only those related to social and economic development, but also those related to welfare contribution and equality generation

Analysing the key factors and drivers of change focusing on the drivers of the social dimension, the scene entry of new generations, e.g. millennials, and the global population ageing, many issues such as mobility and terrorism, entertainment and connectivity, political and religious identity, disability and health, set the impact of aviation sector on social dimension of sustainable development towards 2030 at the top of the agenda.

6 Driving Aviation Social Dimension on Sustainable Development Through Human Development and Welfare

Aviation's social impact on welfare is a major concern for most states, nations, and societies. The complexity and multidimensionality of this social impact address the developing of new and more advanced methods of measuring welfare. This measurement is an essential requirement for the economic and social development of all countries as well as for supporting regional and global cooperation and integration.

Focusing on the aviation welfare impact, efficient, safe, and secure aviation systems offer important access to markets, employment, education, health, and basic services critical to poverty alleviation. At the same time, aviation is a major player in making a big difference in promoting generation of equity and equality (Dimitriou 2017).

Sustainability, based on the concept of the three pillars: environment, social, and economic (Stavins et al. 2003), proposes that sustainable development can only be achieved when each pillar is promoted in concert with the others. On this basis, the defining feature of how sustainability is interpreted is based on the level of welfare and well-being. Increasingly, it is recognized that welfare and its long-term sustainability are the ultimate goals of development and that these notions better capture the human experience of development (Gough and McGregor 2007). The need for wider measures focusing on human well-being has been supported by key stakeholders in the global development arena that have demanded for a multidimensional concept of human well-being.

A sustainability criterion of "well-being" focuses on changes in well-being over time rather than on whether the highest possible level of well-being is achieved at any specific time (Arrow et al. 2004). Within economic frameworks, sustainability is defined to be achieved if the well-being of society is maintained over time (Arrow et al. 2004; Solow 1974).

A different view of the concept of well-being was given to Human Development Report (1990), according to which human development is the main objective of economic development (Sen 1977, 1985). This approach is based on a regulatory framework for the development of the well-being and welfare of the individual and social happiness and evaluates a wide range of people's quality of life such as individual well-being, inequality, and poverty. According to Sen, the human development must target the quality of life and freedom. Addressing human's quality of life as their goal development is the base concept in Human Development Reports (HDR), which focuses on a variety of core functions for the quality of life, such as avoiding death and education.

Based on this human development aspect, several attempts have been made in order to develop more complex indicators for economic developments that are more representative than the GPD per capita. The most recent composite indicator is the Human Development Index (HDI). HDI is an indicator that measures human development, a concept which, according to the United Nations Development Program (UNDP), refers to the extension of individuals desires such as education, health, and income. However, the main goal of HDI is the classification of countries according to the level of human development. This index was created in 1990 by Pakistani economist, ul Haq. HDI is a composite index consisting of three elements, which reflect three dimensions of human development: longevity, knowledge, and access to resources. Life expectancy at birth is selected to measure longevity, knowledge is measured by educational level based on a weighted sum for the rate of adult education and the average number of years of education, and finally, the access to resources is being measured by a real per capita GDP (Rao 1991).

7 Aviation Social Regulation and Protection Principles Towards Sustainable Development

Air transport infrastructures and businesses are subject to a range of regulatory and market forces, which affect the way they forecast and plan their future. Even in markets where there is freedom of entry and exit and where operators are free to set prices and levels of service, regulatory agencies acting in the interest of passengers are trying to ensure that the principles of competitive efficiency and fairness are being complied with. The mixture of market and regulatory forces exists in all sectors of transport operations and impacts to many different aspects (Hensher and Brewer 2001).

The discussion of privatization, interpreted as the sale of assets to the private sector, opens a debate on the gain from exposing business to competition through encouraging competition in the delivery of services by competitive tendering or outright economic deregulation. This raises the question of whether there is more to gain in introducing competition in various ways rather than changing the ownership of the transport business from public to private control.

Since the early 1990s, a dramatic increase in the liberalization of air transport sector and a strengthening of the role of private operators and investors in transport infrastructure worldwide have occurred. Decision-makers and governmental authorities focus on sustained improvement in efficiency and to find additional financing. Indeed, in addition to the important responsibilities of defining policies and strategies, monitoring safety, and financing some of the less attractive segments of the sectors, governments must also now be ready to become fair economic regulators of many of the privately operated transport services and infrastructures.

Shifts in user demands, made choice, technological development, and market forces have increased the need for governments to optimize the air transport system for both public and private sector use. Government regulations generally aim to improve welfare of the public in terms of health, safety, and efficient use of public funds. Regulation and deregulation have significant impacts on competition (Kockelman et al. 2014).

The extent of social regulation has grown over the years as the public has become increasingly alarmed by any number of issues related to the pollution of the environment, employee's safety, and so on. For air transport infrastructure industries without natural monopoly characteristic, the regulator must ensure that competition is promoted and that fair play ensues. The regulator is charged with the task of creating opportunities to enter the market based on managed competitive policy. In this context, air transport sector regulations address a wide range of issues, such as the environment, safety, and employees' wages.

8 Airport Social Dimension on Sustainable Development Key Qualitative Indicators

In order to evaluate the social dimension of transport infrastructure performance on sustainable development, many different quality measures can be specific to each mode of transport infrastructure and to the types of service provided by the mode in a country (OECD 2013). Despite the complexity of the measurements, the level of quality of life is an inherent assumption in decisions about where and how much to invest in transport. It is directly related to management operation as well. The qualitative indicators of airport social dimension are based on a wide range of key non-financial and non-economic value drivers of the air transport infrastructure linked to well-being and human development, and they are defined as follows:

a. **Human Resource Development**

Air transport is an industry where training is critical for addressing compliance requirements in a highly regulated environment. Employee development is key for deploying competitive business strategies and achieving goals. Thus, the target of each airport is to meet training demands, while introducing novel and creative development opportunities based on latest market trends in training and development solutions and technologies. The Human Resource Development indicator is defined as the number of training hours per full-time equivalent.

b. **Social Affect to Local Communities**

Airports are part of a community that extends beyond the infrastructure. The infrastructure operations may impact the neighbours, such as annoyance due to aircraft noise, and thus the goal is to limit, if not eliminate, this impact to the possible extent. This is complemented by targeted investments in social, educational, cultural, athletic, and environmental causes. Therefore, airports incorporate their own initiatives, which aim to improve the quality of life over time. The Social Affect to Local Communities' indicator is the number of actions to be undertaken from the airport within a year in relation to local communities in the areas of:

a. Communication,
b. Infrastructure development,
c. Education,
d. Society,
e. Cultural and athletics, and
f. Environment.

c. **Climate Change Effect**

Capital resources to maintain environmental impact could range from few millions to many billions depending on time and the schedule to mitigate environment implications and the budget allocation in short and long run. While mitigation is mainly focused on actions for short-term impacts, adaptation strategy deals with the industry action plan, including technology innovations to adverse existing trends for global warming and its effects. Therefore, the optimal allocation of capitals in terms of time and activity between adaptation and mitigation is a key question for managers and decision-makers. For most of the cases, it is very difficult to deliver a clear answer (Dimitriou and Sartzetaki 2016). This indicator is measured as the number of actions in the areas of energy, transport, and operations to be undertaken within a year to reduce airport contribution to climate change.

d. **Connectivity/Accessibility Improvement**

Especially when referring to an airport, airport connectivity is the sum of both direct and indirect connectivity from the airport in question—thus measuring the overall level to which an airport is connected to the rest of the world, either by direct flights or by indirect connections via other airports (Dimitriou 2018). Direct connectivity is defined as the air services available from the airport—measured not just in terms of destinations, but also based on the frequency of flights towards the same destination (ACI 2017). Indirect connectivity measures the number of destinations through a connecting flight at hub airports. Thus, the Connectivity indicator measures the overall level to which an airport is connected to the rest of the world, either by direct flights or indirect connections via other airports.

e. **Safety/Security**

Air transport safety is crucial for passengers with safety aspects being usually considered only when accidents occur. However, pertinent regulations and practices ensure the safety of countless passengers on thousands of passenger and cargo transportation. In addition, transport infrastructure engages all operational stakeholders in the implementation of every preventive measure that minimizes the risk of an accident. The Safety/Security indicator is defined as the number of serious (high consequence) incidents/accidents per a certain amount of aircraft movements.

f. **Service Quality**

Optimal management of airport aims to provide quality service and ensure that its staff is available on a 24 hour basis to assist passengers, visitors or users who may need to receive personal assistance or information. The Service Quality indicator is defined as the score from overall satisfaction from air passenger and airport user surveys and is a quantitative research in the form of personal face-to-face interviews.

g. **System Availability**

Airport management aims to create an environment where all passengers may have access to trustworthy sources of technological means and a policy environment that fosters innovation to be able to protect and safeguard their information assets and securely utilize technology. The System Availability indicator is defined as the average availability of the entire transportation system.

9 Methodology Framework for Airports Social Dimension Qualitative Indicators Evaluation

As no single social dimension and qualitative indicator can give a full picture regarding airport performance towards sustainable development, each indicator presents a partial view from a specific viewpoint and is therefore not enough to serve as a basis for management decisions (Marr 2012). The Balanced Scorecard (BSC) is a popular performance measurement scheme suggested by Kaplan and Norton (1992) that was developed to measure performance metrics from financial, customer, and internal processes as well as growth perspectives.

Combining these different perspectives, BSC helps decision-makers to understand the inter-relationships and tradeoffs between alternative performance dimensions, thus leading to improved decision-making and problem-solving (Rajesh et al. 2012). The base point of BSC would be the identification of the unit's strategic plan. This would involve the development of goals, strategy, outputs, measures, and targets.

The performance ratios of the qualitative key social dimension value drivers of the indicators are measured based on this Balance Scorecard framework. Based on the average Score of the Balance Scorecard (ranging between 1 and 5, with 3 being "meet the target") it is assumed that if the score equals with 3 (i.e. "meet the Target"), the performance ratio value is 1. The target values for the year in question are based on an objective rule, i.e. "meet the target" (score 3) was at least the average actual for the previous 5 years while "outstanding" (score 5) was set higher than the best actual value for the previous last years.

10 Conclusions

Air transport development is a crucial issue that planners and decision-makers should be concentrated on for sustainable development. The strategic infrastructure and investments need to include improvements across all major factors that affect the social dimension of sustainable development. It is increasingly recognized that current well-being and its long-term sustainability are the ultimate goals of development and that these notions better capture the human experience of development. The need for wider measures of well-being has been supported by key stakeholders in the global development arena, which have demanded for a multidimensional concept of human well-being quality indicator in order to achieve air transport sustainable development.

This chapter highlighted the challenges in aviation in terms of social benefits diversity. The aviation social benefits towards sustainable development and the respectfulness of human values were defined, and they highlighted the crucial role of aviation's social dimension on sustainable development. In addition, the key qualitative indicators used to evaluate airport's social dimension on sustainable development performance were defined. In doing so, a tool was provided for airport management to monitor the airport's social dimension on sustainable development performance. Conventional wisdom is to stimulate the interest on topic and promote key actions towards aviation and especially airport social dimension on sustainability.

References

ACI. (2017). *Airport Industry Connectivity Report 2017* (pp. 3, 9–11). https://www.aci-europe.org/policy/connectivity.html.

Arrow, K., Dasgupta, P., Goulder, L., Daily, G., Ehrlich, P., Heal, G., et al. (2004). Are we consuming too much? *Journal of Economic Perspectives, 18* (3), 147–172.

Connelly, S. (2007). Mapping sustainable development as a contested concept. *Local Environment, 12*(3), 259–278.

Dasgupta, S., Hemamala, H., & Wheeler, D. (2000). What improves environmental compliance? Evidence from Mexican Industry. *Journal of Environmental Economics and Management, 39*(1), 39–66.

Dimitriou, D. (2016). *Climate change implications in transport and tourism market equilibrium, climate change management* (pp. 409–424). Cham: Springer.

Dimitriou, D. (2017). *Air transport economic footprint in remote tourist destinations, book: Mobilities, tourism and travel behaviour.* InTech. ISBN 978-953-51-5608-6.

Dimitriou, D. (2018). Evaluation of tourist airports productivity towards tourism development. *Journal of Cogent Business and Management.* https://doi.org/10.1080/23311975.2018.1464378.

Dimitriou, D., & Sartzetaki, M. (2016). Sustainable development variables to assess transport infrastructure in remote destinations. *International Journal of Urban and Civil Engineering, 10*(10), 1343–1350.

Dimitriou, D., & Sartzetaki, M. (2017). Air transport connectivity development in tourist regions (Working Papers SIET 2018). ISSN 1973-3208.

Dimitriou, D., & Sartzetaki, M. (2018a). Assessing air connectivity socio-economic footprint. *International Journal of Transportation Science and Technology, 7*(4), 283–290.

Dimitriou, D., & Sartzetaki, M. (2018b). Social dimension of air transport sustainable development, world academy of science, engineering and technology. *International Journal of Industrial and Business Engineering, 12*(4), 568–571.

Dimitriou, D., Voskaki, A., & Sartzetaki, M. (2014). Airports environmental management: Results from the evaluation of European airports environmental plans. *International Journal of Information Systems and Supply Chain Management (IJISSCM), 7*(1), 1–14.

Dimitriou, D., Mourmouris, J., & Sartzetaki, M. (2017). Quantification of the air transport industry socio-economic impact on regions heavily depended on tourism. *Transportation Research Procedia, 25*(2017), 5242–5254.

EC—European Communities. (1992). *Treaty on European Union, treaty of Maastricht*. Luxembourg: Official Journal of the European Union C 191 of July 29, 1992.

EC—European Communities. (1997). *Treaty of Amsterdam amending the treaty on European Union, the treaties establishing the European communities and certain related acts*. Luxembourg: Office for Official Publications of the European Communities. ISBN 92828-1652-4.

Gough, I., & McGregor J. A. (2007). *Well-being in developing countries: New approaches and research strategies*. Cambridge University Press.

Hazakis, K. (2017). An institutional proposal for economic adjustment in euro area. *Journal of Sustainable Development, Culture, Traditions, 1*, 82–100.

Hensher, D. A., & Brewer, A. M. (2001). *Transport: An economics and management perspective* (1st ed.). New York: Oxford University Press.

IATA. (2017). *Future of airline industry 2035*. https://www.iata.org/policy/iata-future-airline-industry.

ICAO. (2017). *Aviation benefits beyond borders*. https://www.icao.int/sustainability/Documents/AVIATION-BENEFITS-2017.

ITF. (2017). *ITF Transport Outlook 2017*. Paris: OECD Publishing. https://doi.org/10.1787/9789282108000-en.

Jacobs, M. (1999). Sustainable development as a contested concept. In A. Dobson (Ed.), *Fairness and futurity: Essays on environmental sustainability and social justice* (pp. 21–45). Oxford, UK: Oxford University Press.

Kaplan, R. S., & Norton, D. P. (1992, January–February 6). The balanced scorecard as a strategic management system. *Harvard Business Review*, 1–66.

Kockelman, K. Chen, D., Larsen, K. A., & Nichols, B. G. (2014). *The economics of transportation systems: A reference for practitioners* (Paperback—United States).

Lucas, R. (1988). On the mechanics of economic development. *Journal of Monetary Economics, 22,* 3–42.

Léle. (1991). Sustainable development: A critical review. *World Development, 19,* 607–621 (ÛÂÏ).

Marr, B. (2012). *Key Performance Indicators (KPI): The 75 measures every manager needs to know* (1st ed.) (Financial Times Series).

OECD. (2011). *Perspectives on global development 2012: Social cohesion in a shifting world.* Paris: OECD Publishing. http://dx.doi.org/10.1787/persp_glob_dev-2012-en.

OECD. (2013). *How's life? 2013: Measuring well-being.* Paris: OECD Publishing. http://dx.doi.org/10.1787/9789264201392-en.

Rajesh, R., Pugazhendhi, S., Ganesh, K., Yves Ducq, & Lenny Koh, S. C. (2012). Generic balanced scorecard framework for third party logistics service provider. *International Journal of Production Economics, 140*(2012), 269–282.

Rao, V. V. B. (1991). Human development report 1990: Review and assessment. *World Development, 19*(10), 1451–1460.

Samberg, S., Bassok, A., & Holman, S. (2012). Method for evaluation of sustainable transportation. *Transportation Research Record: Journal of the Transportation Research Board, 2242,* 1–8.

Sen, A. (1977). Social choice theory: A re-examination. *Econometrica, 45*(1), 53–89.

Sen, A. (1985). Well-being, agency and freedom: The Dewey lectures 1984. *Journal of Philosophy, 82*(4), 169–221.

Sen, A. (2000). *On the mechanics of economic development* (Robert E. LUCAS, Jr.). Chicago, IL: University of Chicago. Received August 1987, final version received February 1988 Development as Freedom. New York: Anchor Books.

Solow, R. (1974). Intergenerational equity and exhaustible resources. *Review of Economic Studies, 41,* 29–45.

Stavins, R., Wagner, A. F., & Wagner, G. (2003). Interpreting sustainability in economic terms: Dynamic efficiently plus intergenerational equity. *Economics Letters, 79,* 339–343.

UNDP. (2014). *Human development report 2014—Sustaining human progress: Reducing vulnerabilities and building resilience.* New York, NY: United Nations Development Programme. http://hdr.undp.org/sites/default/files/hdr14-report-en-1.

UNECE. (2013). Chapter 2: Climate change implications for transport. In *Climate change impacts and adaptation for international transport networks* (pp. 19–46). Geneva: United Nations Economic Commission for Europe.

UNECE. (2015). *Transforming our world the 2030 sustainable development agenda*. A/RES/70/1. http://www.un.org/sustainabledevelopment/sustainable-development-goals.

UN—United Nations. (1992, June 3–14). *General assembly, report of the United Nations conference on environment and development, Rio de Janeiro*. New York, NY: United Nations Publications, A/CONF.151/26 (Vol. I).

UN—United Nations. (1993). *Agenda 21: Earth Summit—The United Nations programme of action from Rio* (294pp.). New York, NY: United Nations Publications. ISBN 13: 9789211005097.

UN—United Nations. (2002, August 26–September 4). *Report of the world summit on sustainable development, Johannesburg, South Africa*. New York, NY: United Nations Publications, A/CONF.199/20. ISBN 92-1-104521-5.

UN—United Nations. (2006). *The millennium development goals report*. New York, NY: UN.

UN—United Nations. (2007). *Indicators of Sustainable Development: Guidelines and Methodologies* (3rd ed.). New York, NY: United Nations Department of Economic and Social Affairs, United Nations Publication. ISBN 978-92-1-104577-2.

World Bank. (2013). *World development indicators* (Report). Washington, DC: World Bank.

CHAPTER 10

An Element-by-Element Approach for a Holistic Estimation of the Airport Carbon Footprint

Maria Nadia Postorino and Luca Mantecchini

1 Introduction

Since 1970, there has been an increasing growth of worldwide aviation industry—on average by 5% yearly. Emissions from civil aviation have increased accordingly, and greenhouse gas (GHG) emissions from international aviation have grown by 87% since 1990. According to some studies (IATA 2016; Boeing 2016; Airbus 2017), air traffic will continue growing in the next years at an annual rate of 4–5%. Such expected growth has strengthened the attention of stakeholders on the environmental carbon effects that developing airport traffic could generate locally. Particularly, ICAO recognized the importance of the airport carbon footprint since 2007, when the project "Carbon emission calculator," successfully launched in 2008, started and was then followed

M. N. Postorino
University of Reggio Calabria, Reggio Calabria, Italy
e-mail: npostorino@unirc.it

L. Mantecchini (✉)
DICAM, University of Bologna, Bologna, Italy
e-mail: luca.mantecchini@unibo.it

T. Walker et al. (eds.), *Sustainable Aviation*,
https://doi.org/10.1007/978-3-030-28661-3_10

by other reports (e.g., ICAO 2011, 2016). In this perspective, carbon impacts produced by airport activities have received increasing attention in the last years (see, e.g., Yu et al. 2004; Lu and Morrell 2006; Graham 2014; Ashok et al. 2017).

For their part, airport operators are adopting several measures addressed to achieve the ACI Airport Carbon Accreditation, which is a four-step process where carbon-related airport operations are identified, their impacts mapped, and then management and engagement plans are identified to manage, reduce and neutralize them. The effectiveness of the plans is strongly related to the implemented actions, which in turn depend on their capability to reduce—and possibly remove—the effects of relevant carbon sources. The main aim of the program is to promote the implementation of airport best practices for carbon and energy management, in order to reduce and counterweigh CO_2 emissions. The program deals also with the improvement of ground access transportation systems, the processes of production and consumption of energy in terminal buildings and the activities performed in the airside. It is addressed to promote a close cooperation among airport stakeholders, research institutions and public authorities. At European level, currently 136 airports are involved in this program, covering 92.3% of EU air passenger traffic (Airport Council International 2018). In addition, 39 airports have reached the carbon neutrality status for the activities under their direct control.

According to the annual ACI carbon accreditation report (year 2017), great progresses have been made compared to 2016. The most positive figure is that related to the increasing number of airports that have joined the program. The number of member airports increased by 21% from the previous year, thus covering almost 40% of the global traffic on scheduled flights.

As introduced above, the four-step certification process establishes four levels of certification—Mapping, Reduction, Optimization and Neutrality—evaluates the commitment of each airport and the actions taken to cut emissions and classifies them accordingly. The first level of certification takes place when the airport is mapping its emissions and developing a plan to manage them, while the most advanced certification proves that the structure is using every resource to reduce its environmental impacts.

More in detail, level 1 (Mapping) requires airports to produce a complete carbon footprint estimate certified by an external evaluator along with evidence of a carbon policy endorsed at the highest levels of airport management. The estimated carbon footprint serves as basis for implementing carbon management and engagement plans (level 2 Reduction and level 3 Optimization). Airports can finally achieve CO_2 neutrality by totally compensating their emissions (level 3+: Neutrality).

All over the world, 48 airports have achieved the ambitious goal of "Level 3+Neutrality" (ACI Europe 2018). Among these, only one (Dallas/Fort Worth, DFW) is in North America and 38 are in Europe (10 in Sweden, 6 in Italy, 4 in France and UK). Such recognition is the result of decades of efforts and commitments to reduce carbon emissions, to help clean up the air and to limit environmental impact in the airport neighborhood.

The "zero emissions" goal is achieved when the net emissions of carbon dioxide for a whole year are zero, i.e., when the airport absorbs or compensates the same amount of CO_2 it produces. Since 2010, DFW has achieved a 29% reduction in carbon emissions on a per-passenger basis, as well as an overall reduction of 38% in energy costs, while total passengers at the airport have increased by 15% over the same period.

In Europe, Stockholm Arlanda Airport achieved Level 3+Neutrality cutting its CO_2 emissions by more than half over the past seven years (ACI 2018). The airport buildings are warmed up with heating from biofuel, and all electricity supplied is produced from renewable sources. According to Manchester Airports Group (MAG), two airports—East Midland and Bournemouth—for the past six years have implemented some actions to cut down CO_2 emissions and reach their carbon neutral accreditation. In Italy, Treviso Airport received its Level 3+Neutrality certificate and Venice Airport has been at the highest level of Airport Carbon Accreditation since 2014. More generally, during the period 2017–2018 the 133 accredited airports in Europe succeeded at reducing the CO_2 emissions under their direct control by 163,277 tons, compared to their average emissions of the three previous years (ACI 2018).

The airport's interests to receive their certification are increasing all over the world. However, best practices and results often depend on local contexts and airport capability to identify effective actions.

Starting from this brief overview, the identification of carbon emission sources and their related impacts is of a great relevance to set suitable airport policies. Particularly, addressing the most significant emission sources allows finding more efficient solutions and identifying more effective policies, as some studies already suggested (e.g., Morrinson 2009; Liebert and Niemeier 2010; Postorino and Mantecchini 2014; Postorino et al. 2016).

In line with the previous considerations, this chapter proposes a methodological framework to identify effective policy actions based on the segmentation of carbon emission sources at the airport. In other words, the relevant airport activity (such as airside, terminal and landside operations) contributing to the airport carbon impact is split into elementary segments. In turn, each elementary segment is described by a vector of "factors," which represent quantitative and/or qualitative features of the segment. The environmental effects produced by each elementary segment—which is a direct or indirect source of pollution—are computed, and then actions that could be implemented to reduce or possibly remove the produced environmental impacts are identified (Heinrich et al. 2014; Postorino et al. 2019).

2 Description of the Element-by-Element Approach

When dealing with the description of a system, particularly a complex system, this latter is often split into a first logical level of components that have similar elements. In turn, each of these is further split into subsystems up to the lowest hierarchical level where the "elementary" component may be found, which can be part of one or more subsystems.

To be effective, splitting the project into operational packages requires that such partition is complete and includes every detail. The partition of the project into smaller units is the basis for a more detailed planning of resources and budget among the others.

When progressing top-down in the exploration of complex systems, the problem that often arises is the identification of the "elementary process," i.e., the level such that it is no longer possible to proceed with the process decomposition. A general rule that may be pursued is the following: The "elementary process" is the decomposition level such that the identified element within it is clearly limited and self-contained. In other words, the "elementary segment" is at the beginning of a sequence of more or less complex processes that produce finite and well-identified outputs, which in turn will be used for another process or will be used outside the context under consideration.

Such functional decomposition is a method of analysis and representation based on the aggregation of elements or components of the examined system with common logical-functional characteristics. This general approach is used in the field of technical-scientific disciplines to describe in a schematic and systematic way, even in the smallest details, complex systems, in both structural and functional aspects.

The decomposition procedure is not necessarily unique as it depends on the relevant question: "What do we want to achieve?" In other words, decomposition is related to the objective.

The application of such method to the airport context—which is a complex system—requires the identification of the components with similar elements. Such components correspond to the "airport activities"—such as ground airport access, landing and take-off, which are all transport-addressed activities.

In this perspective, the element-by-element (EbE) approach discussed in this chapter follows a two-step method (Fig. 1) First, the airport activities contributing to the carbon impacts are identified within a general framework, which tries to represent the several facets of the airport. Second, the activity is analyzed by representing logical level of components—as in the description above—that are split into elementary segments (E), each one of them depending on factors (F) that characterize E. Then, for a given activity A_l, each E_i is a function of a vector $\underline{F}$ and the output O_l it produces (carbon emission in this context) depends on the values assumed by the components F_j of $\underline{F}$.

Then, for a given A_l:

$$E_i = f(F); \quad F = \{F_j, j = 1, \ldots, m\} \quad O_l = g\left(E_i = f\left(F_j, j = 1, \ldots, m\right)\right)$$

The identification of the factor(s) for each segment E_i is an important step of the decomposition process, because in the airport context they are at the beginning of a sequence of other processes producing well-identified outputs, particularly carbon impacts for the aim of this study. Factors F_j giving the higher contribution to the airport carbon footprint may be identified, and then targeted policies/actions inspired by an efficiency-based concept may be adopted for the corresponding element E_i.

The advantage of such an approach is the identification of the elementary segment and the related factors that describe the segment itself. In fact, for a given airport activity A_l the output $O_{l,i}$ produced by segment E_i depends on quantitative and/or qualitative values assumed by factors F_j of E_i. Then, acting on F_j is more efficient than acting generically on the whole segment E_i and/or activity A_l.

It is worthwhile to note that the identification of the elementary segment, and then the corresponding factors, can be made at several levels of disaggregation, also depending on aims and scopes as well as on the relevance of elementary segments for the actors applying the

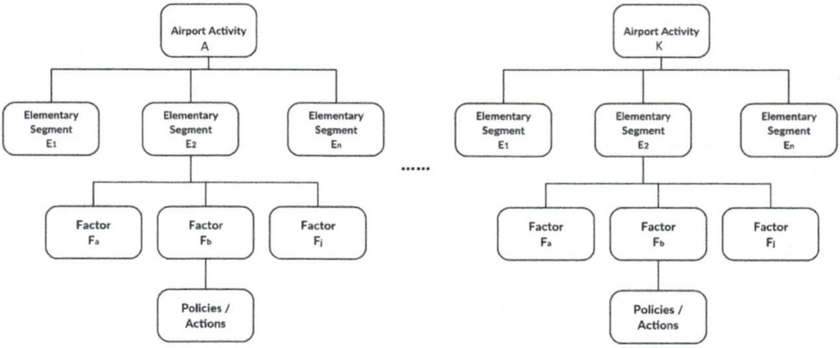

(a) General framework

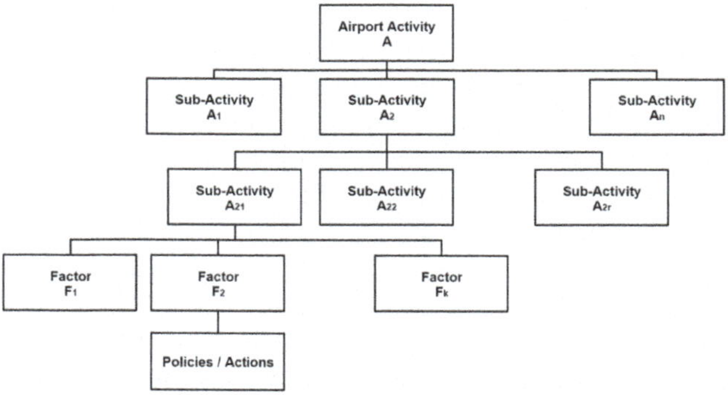

(b) Several decomposition levels of the Element-by-Element framework

Fig. 1 The element-by-element framework to identify effective actions

approach. In other words, the elementary segment is the disaggregation level relevant to the identification of effective actions. In cases of complex activities, they can be split into simpler parts (sub-activities) corresponding to a first level of segmentation—still depending on aggregate factors. In turn, each part will be split into other ones until what is intended as "elementary" segment is found. By following this decomposition process, relevant factors may be identified and effective actions may be set. Analysis at intermediate levels—i.e., before reaching elementary segments and corresponding relevant factors—may provide a preliminary assessment of the impacts associated with a given activity. Figure 1b summarizes these further details of the EbE approach.

3 Airport Transport Function and Related Activities Generating Carbon Emissions

As briefly shown in the previous section, the airport carbon footprint is the result of several airport and airport-related activities that can be split into elementary segments each one of them influenced by several factors, which in turn produce carbon emissions (outputs). From this perspective, the first step is the identification of such activities and their carbon impacts, together with actions that airport operators (but also other interested stakeholders such as regulator and/or airlines where it is the case) might adopt to reduce it.

It is well known that airport is complex nodes where more activities take place. From one side, they are transport terminals, whose only goal is the management of traffic flows—passengers and aircraft—in order to realize safe and efficient interchanges from ground modes to air mode—and vice versa.

It is worthwhile to note that the carbon impacts discussed here are the ones produced by the airport transport function, to which two main transport variables (aircraft movements and air transportation demand in terms of both passengers and goods) are associated. In fact, airport primary function is "transport," i.e., goods and passengers carried from one place to another by using aircraft. Following the schematic representation depicted in Fig. 2, an airport can be seen as an inter-modal transport node, where a continuous flow of arriving passengers is transformed in a discrete process of departing aircraft and vice versa.

More in detail, the flow of passengers starts from/arrives at the airport catchment area *that can be defined as* the geographic *basin* of influence of a given airport. The catchment area influences directly the market power of an airport, and it is *defined* by several variables, including quality and capacity of ground access transportation systems, number of air destination served, flight frequency and scheduling, level of service offered and other facilities. The air demand accessing the airport is usually estimated for each available transportation mode, by direct surveys or by stage demand models—mainly within a discrete choice approach. Such models allow characterizing the travel demand for several trip features, such as the origin and/or destination airports, the ground access transport modes, the air services at airports (see, e.g., Pels et al. 2003; Suzuki 2007; Jou et al. 2011). At the end, the share of users on available ground access transport modes may be obtained by using random

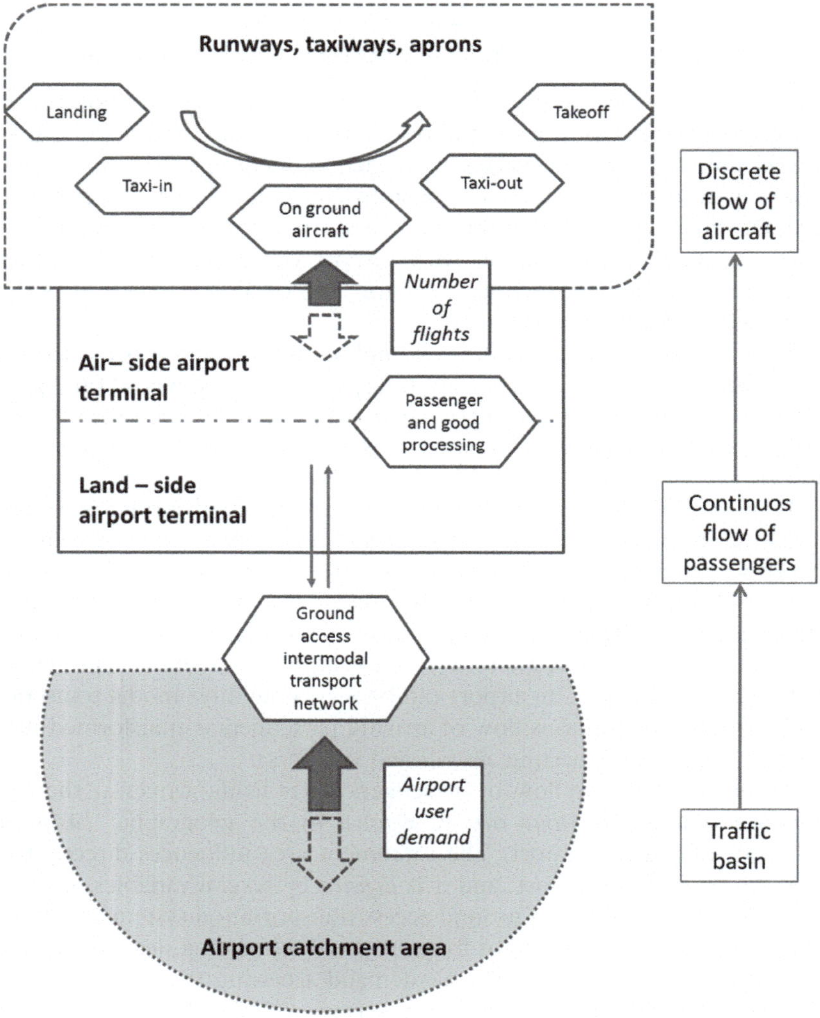

Fig. 2 Summary of the airport transport function

utility models such as Logit, Nested Logit and Cross-Nested Logit models (Train 2003).

From an operational point of view, usually an airport is divided into two areas: landside and airside. Landside facilities allow satisfying passenger needs and processing goods and baggage. The most relevant activities are related to the effective movement, processing and control of the flow of passengers, baggage and goods at terminal buildings, in order to meet passengers' expectations concerning the level of service in the whole airport and then improve passengers' experience. The landside is usually composed by a main terminal building, cargo terminal, car parking and other facilities designed to allow and improve ground access. In the airside, the main processes taking place involve aircraft operations during LTO cycle (landing, takeoff, taxi-in and taxi-out) and ground handling services (such as pushback, baggage handling, cleaning, passenger service, catering, de-icing and refueling). To perform both landside and airside activities described above, a great amount of energy is necessary. Among the most widely used energy sources, electricity is the dominant one at airports, even if during the last years it is possible to find in the literature the use of other kinds of energy sources such as renewable energy technologies (Kılkış 2014; de Santoli et al. 2015).

By considering the airport transport function, the carbon emission due to both energy consumption and traffic can be linked to the identified relevant transport variables, i.e., passengers and aircraft movements. In the landside, the energy consumption and the related emissions for terminal operations (considering also the ground access transport systems used by passengers to move to/from the airport) are directly linked to the expected number of passengers at the airport. In fact, terminal facilities (such as check-in and security controls, baggage facilities, restaurants, retail and all the spaces necessary for passenger processing) require energy and then can be considered sources of carbon emissions according to the way such energy is produced (Ryley et al. 2013; Huang et al. 2015). At the same time, ground access transportation systems generate relevant amount of emissions (Budd et al. 2016).

On the airport airside, the main sources of air pollution—directly linked to aircraft movements—are exhaust from aircraft engines (main engines and auxiliary power units), diesel engines (from handling vehicles and other equipment, whose number and characteristics are related to the number of movements), direct fuel evaporation from refueling and particles from vehicles and aircraft brakes, tires and asphalt cleavage. It is worthwhile to note that the number of aircraft movements at a given airport depends on the actual air transportation demand and on the average break-even load factor.

The most important aircraft emissions are carbon dioxide (CO_2), nitrogen and sulfur oxides (NO_x and SO_x) and particulate matter (PM). More in detail, aircraft emissions in the airside are related to both the number of movements and the level of technology and age of the operating fleet mix. It is widely recognized that new aircraft is 70% more fuel-efficient than they were 40 years ago (Babikian et al. 2002; Zou et al. 2014). Furthermore, carbon monoxide emissions have been reduced by 50%, while unburned hydrocarbon and smoke have been cut by 90% (Graham et al. 2014).

Aircraft emissions can be estimated on the basis of their main airside operations, i.e., landing/take-off and taxiing. Aircraft-related emissions during landing and take-off depend mainly on the aircraft characteristics (aircraft size, age, weight and fuel efficiency), while for taxiing they depend mainly on the taxiway configuration (as path length, presence of high-speed taxiways) and on taxiing technologies and procedures (Postorino et al. 2019). Finally, further airside emissions depend on extra fuel burnt caused by airport congestion (Kuhn 2013).

The actual amount of airport emissions varies significantly from airport to airport, as it depends on passengers and movements at the examined airport. Then, it can be quantified at different levels—from average contributions depending on the average source characteristics, to the detailed simulation of the whole passenger journey to/from the airport or the microscopic representation of every stage of ground operations and their related emissions.

Following the airport transport function perspective considered here, and as depicted in Fig. 3, the airport carbon footprint estimate can be obtained by considering specific and unitary contributions deriving from processed passengers and/or aircraft movements.

More in detail, the identification of carbon emission (macro-)sources linked to the relevant transport variables—i.e., passenger demand and number of aircraft movements within the same reference period—and to specific airport activities allows obtaining separate carbon emissions, which together contribute to the airport carbon impact. In line with the EbE approach, the identification of transport-related airport activities

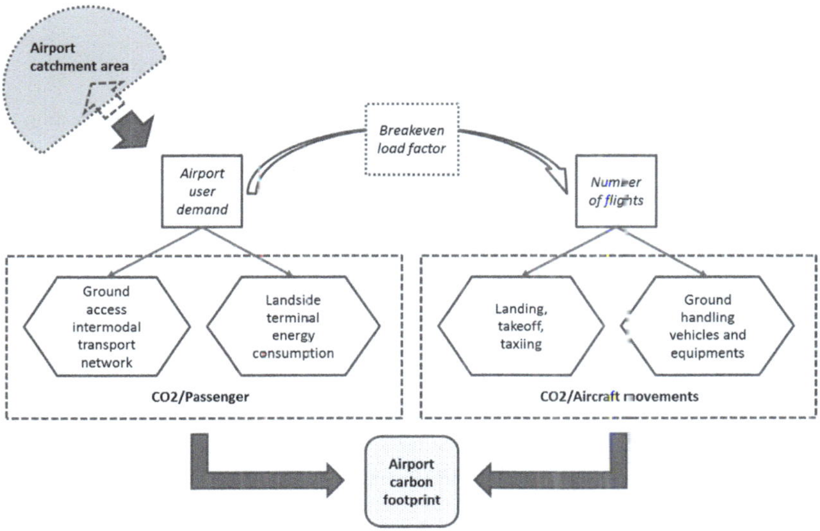

Fig. 3 Airport carbon footprint and transport contributions

and their (macro-)sources is the first step to identify targeted actions addressed to achieve carbon neutrality.

For example, as reported in Fig. 3 relevant macro-sources of carbon emissions linked to passengers are ground access transport systems and landside terminal energy production and consumption, while macro-sources linked to aircraft movements are LTO aircraft and ground handling.

According to ACI (2009), airport CO_2 emissions sources can be classified in "scopes," as reported in Table 1. More in detail, emissions from ground access transport systems can be identified in terms of Scope 3, on-ground LTO aircraft emissions and handling vehicles could be still identified as Scope 3, even if airports can take actions and measures to control and reduce them (e.g., effective taxiway and taxiway exits design), while energy produced/consumed for airport terminal landside activities can be identified in terms of Scope 1 and 2. This classification, and in particular the distinction between sources under or out the direct control of airports, aims to promote cooperation strategies among the stakeholders (airport owners, planners, local/national authorities) involved in the development and planning of airport infrastructure, as the final and common goal is the achievement of sustainability and an improved quality of life.

Table 1 ACI classification of airport GHG emissions

Scope 1	GHG emissions from sources that are owned or controlled by the airport operator
Scope 2	GHG emissions from the off-site generation of electricity (and heating or cooling) purchased by the airport operator
Scope 3	GHG emissions from airport-related activities from sources not owned or controlled by the airport operator *Scope 3A*: emissions which an airport operator can influence, even though it does not control the sources *Scope 3B*: emissions which an airport operator cannot influence to any reasonable extent

Source ACI (2009)

4 Application of the EbE Approach to Aircraft-Related Activities

Following the EbE approach, some selected airport activities have been split into elementary segments and the corresponding carbon contribution has been computed in order to better identify the related environmental effects. Particularly, two test cases have been considered, both due to (macro-)sources linked to aircraft movements. The first one refers to taxi-out operations and the second one to handling activities.

In the first case, the taxi-out operation has been split into elementary segments, with a relative high level of decomposition. The second activity, on the contrary, has been considered at a lower level of decomposition; particularly, the main activity (handling) has been split into relevant parts, each one characterized by one or more factors, but without a relevant decomposition in "elementary segments." The aim is to show two different ways to apply the EbE approach. The first one may allow identifying effective actions to reduce the airport carbon impacts linked to taxi-out operations. The second one provides a first assessment of the impacts caused by several sub-activities related to handling operations, without reaching significant decomposition levels that may help airport operators to identify targeted actions. However, such intermediate results provide an initial estimate of the impacts caused by handling operations and allow identification of the sub-activities deserving in-depth analyses.

Figure 4 summarizes some relevant airport activities due to the airport transport function. The figure also reports some possible strategies, some of them usually adopted by/suggested to stakeholders.

By focusing on one of these activities, it may be split into elementary segments, which in turn depend on one or more relevant factors. Simulation of strategies by acting on one or more factors helps the

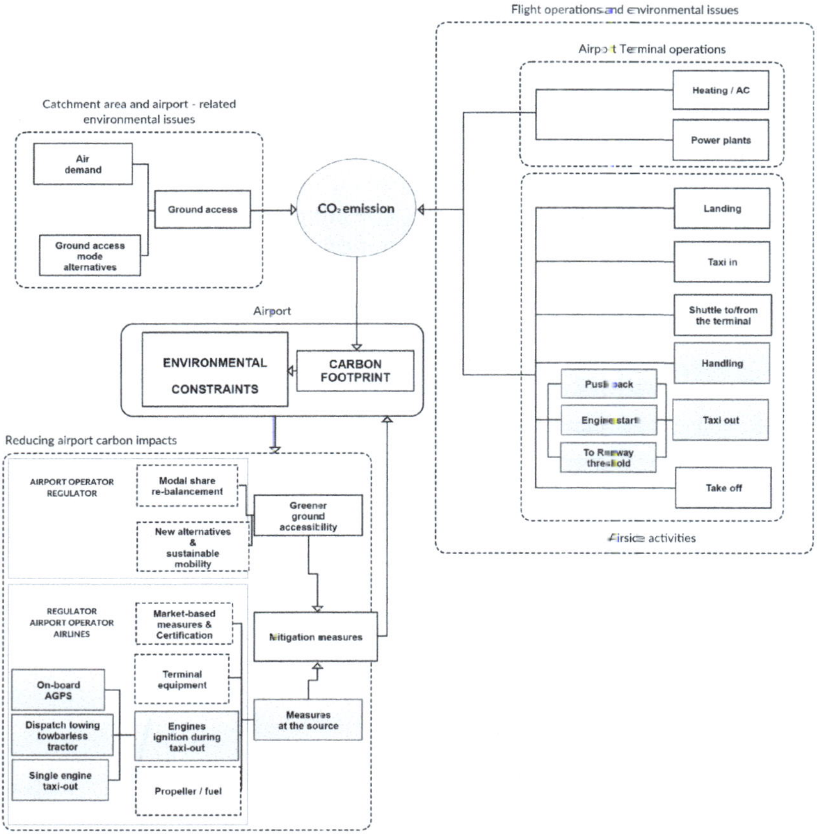

Fig. 4 Relevant airport activities due to the airport transport function and commonly adopted strategies (*Source* Postorino et al. 2019)

airport operators to select effective actions that may reduce significantly the airport carbon footprint.

As an example, taxiing procedures have been split into three elementary segments E_i (pushback, engine start, taxiing out to line-up). More in detail, following the EbE approach depicted in Fig. 1 the activity A= "taxi-out" has been split into E_1 (pushback), E_2 (engine start and warm-up) and E_3 (taxiing out to line-up). During the "pushback" segment, a suitable tug tows the aircraft away from the gate. The pushback phase occurs when the aircraft is not allowed (or not able) to self-maneuver. After, the "engine start" segment corresponds to the engine warm-up. Finally, the "taxiing out to line-up" segment completes the taxi-out procedure along the taxiway up to the runway (Fig. 5).

For each segment E_i, the relevant factors F_j contributing to the carbon taxiing emissions have been identified in the time spent along the segment and the corresponding fuel consumption. In fact, the amount of carbon emission depends on the duration of each operation along the corresponding elementary segment.

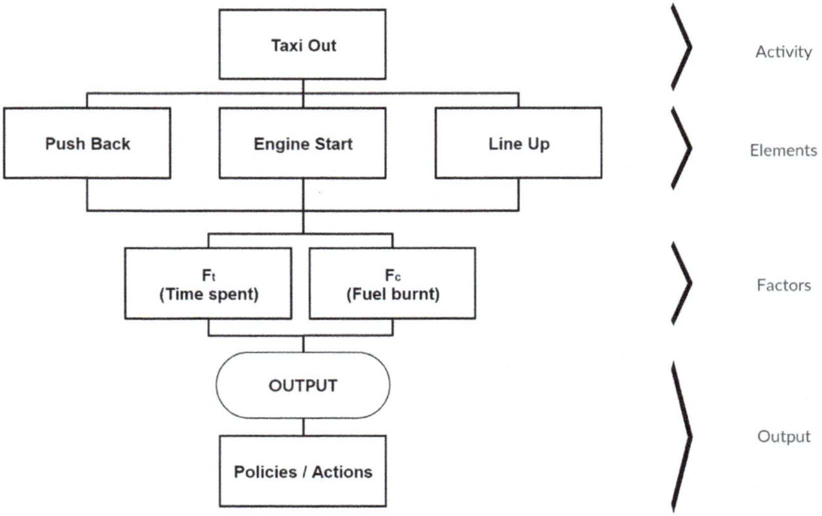

Fig. 5 Application of the EbE approach to taxing-out activities

More in details, and as depicted in Fig. 5, the relevant factors F_t (time spent along each elementary segment) and F_c (fuel burnt in liters per time unit) provide the output O expressed as:

$$\underline{O} = g(\underline{E}(\underline{F})) \quad \underline{E} = \begin{Bmatrix} E_1 \\ E_2 \\ E_3 \end{Bmatrix}; \quad \underline{F} = \begin{Bmatrix} F_t \\ F_c \end{Bmatrix}; \tag{1}$$

The actual output O can be computed by making explicit the function $g(.)$. In this case, the total amount of carbon dioxide emitted during the taxi-out phase by "n" aircraft movement has been expressed as:

$$O = \sum_{i=1}^{n} \left(F_{t1,i} \cdot F_{c1,i} \cdot \lambda_P + F_{t2,i} \cdot F_{c2,i} \cdot_{A,i} + F_{t3,i} \cdot F_{c3,i} \cdot \lambda_{A,i} \right) \tag{2}$$

where $\lambda_{(.)}$ is the specific emission factor, expressed in grams of pollutant per liter of fuel burnt (for both aircraft, λ_A, and pushback vehicles, λ_P) and $F_{(.)j,i}$ are the considered factors $F_{(.)}$ (time and fuel burnt) associated with the three segments (1, 2 and 3) for each aircraft i. As for $\lambda_{(.)}$, the estimation of specific emission factors due to several pollution sources has been widely studied in the literature. Particularly, the reference unit emission factors due to airport sources—such as handling vehicles and aircraft—could be obtained by several databank (Greenhouse Gas Protocol Italian ISPRA, Italian Greenhouse Gas Inventory 1990–2010; ACRP Report 11 2009; Airport Air Quality Guidance Manual ICAO Doc 9889 2011; ICAO Aircraft Engine Emissions Databank 2018).

Finally, the products $(F_{tj,i} \cdot F_{cj,i} \cdot \lambda_{(.)})$ in Eq. (2) provide the contribution of each segment to the whole carbon output due to the taxi-out activity.

The second example discussed here refers to handling activities, represented—as previously introduced—at a lower level of decomposition. Particularly, the focus is on the several "sub-activities" in which the main activity can be divided, each one characterized by other sub-activities as depicted in Fig. 1b.

Figure 6 shows the several parts in which the macroactivity "handling"—which here includes also the turnaround of the related equipment—has been decomposed.

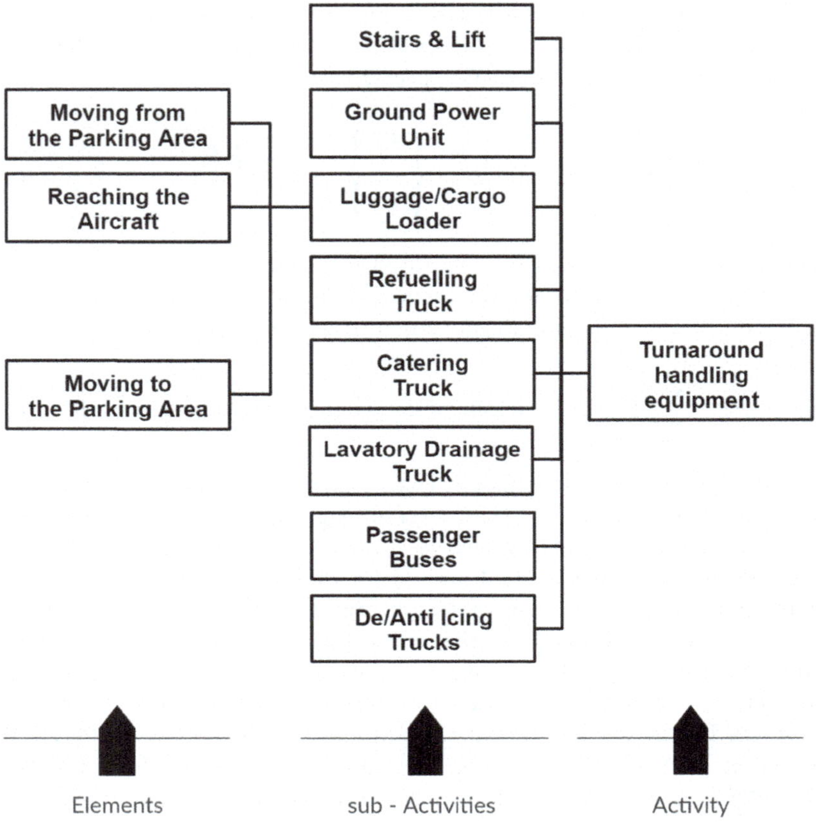

Fig. 6 Decomposition of handling activities in more sub-activities

Following the same approach adopted in the previous example, each sub-activity $A_{(.)}$ contributes to the carbon emission similarly to the elementary segment E. In this case, however, the sub-activity contributes as a whole; in other words, it corresponds to an aggregate element, which may be decomposed in further sub-activities until the "element"—from the point of view of the operator—has been reached. As an example, the "Luggage/Cargo loader" sub-activity may be further split into several elements (as depicted in Fig. 6).

The factors associated with the sub-activity $A_{(.)}$ may be still the time spent, F_t, to realize it and the associated fuel burnt, F_c—if the use of a vehicle is required for the corresponding segment, as in the case of "Luggage/Cargo loader." However, the factors associated with the sub-activity depend on its specific features, although linked to the production of carbon emissions for the goal of this study. As in Eq. (2), the average emissions O associated with n aircraft movements have been expressed as:

$$O = \sum_{i=1}^{n} \left(F_{t1,i} \cdot F_{c1,i} + F_{t2,i} \cdot F_{c2,i} + \cdots + F_{tm,i} \cdot F_{cm,i}\right) \cdot \lambda_H \quad (3)$$

where λ_H is the average, specific emission factor for handling vehicles and subscript "m" refers to the m sub-activities in which the handling activity has been decomposed.

The main difference between Eqs. (2) and (3) is that Eq. (2) refers to elementary segments of the activity (or sub-activity) $A_{(.)}$, while Eq. (3) refers to sub-activities, which correspond to aggregate elements. At the same time, the factors F_t and F_c in Eq. (3) are referred to average values (in time and fuel consumption, respectively) of each sub-activity and not to an elementary segment. Then, Eq. (3) provides a preliminary estimate of the impact generated by a given (sub-activity), while Eq. (2) other than providing a more detailed estimation of the impact allows the airport manager to act directly on the elementary segment to which the higher emission levels are associated.

5 Discussion and Conclusions

The identification of "elements" is a typical engineering process, in particular adopted by that part of the engineering field studying complex systems. System engineering studies the way in which several elements interact each other by providing a final state of the system that depends on the interdependence relationships among several "parts." These latter may be identified in elementary parts or aggregate parts, according to the level of details and the level of segmentation useful for practical applications.

However, such identification is often adopted for macro subsystems generally corresponding to processes or complex artificial systems. So far, this basic principle has not been adopted to decompose airport transport "activities" as in the approach proposed here. Furthermore, while often the goal is to aggregate elements in order to represent the complex system, here the approach has been reversed. In fact, the activity (or generally the process) has been decomposed until an "element" is reached. The goal is to identify which element contributes to a given output in a more relevant way.

The output is here the impact produced by airport activities—and specifically carbon impacts. The application of the EbE approach as described and applied in the previous sections has shown the potentialities of the method. It is worthwhile to note that it was not the aim of this study to propose specific computations of carbon emissions by applying the EbE approach (interested readers may refer to Postorino et al. [2019], for a specific test case). In fact, the focus is on the method in itself and the opportunity it offers to identify targeted, effective actions.

The EbE approach allows airport managers to act on the factors F_j influencing each individual element E_i in order to optimize the process efficiency and reduce impacts. As an example, two possible applications of the method to airside airport activities generating carbon impacts—a first application with higher decomposition level and a second one with only a first, aggregate level of decomposition—have been described.

In the example of the taxi-out procedure, it is possible to act on the "push-back" and "warm-up" segments by adopting self-guided or less-pollutant vehicles (i.e., electric ones), postponing the ignition of aircraft engines or, again, taxiing out with only one engine on. In fact, the carbon contribution for taxi-out operations is due to both aircraft and pushback vehicles, if they need. For both, their carbon contribution is linked to the fuel burnt during the operating time associated with each elementary segment—i.e., pushback, engine start and warm-up, taxiing out to line-up. By modifying the time associated with each segment for each moving vehicle (tug or aircraft), the whole carbon contribution—depending on fuel burnt—will change consequently according to the specific emission factors. Particularly, such factor is higher for aircraft than for tugs. Then, although an increase in tug operating time corresponds to higher fuel consumption for the related segments, however, the reduction of aircraft engine functioning time produces a more than linear reduction of carbon emissions along the corresponding segments.

Equation (2) allows computing the output (carbon emissions) for several combinations of operating times for aircraft and tugs, according to taxi-out procedures. As discussed in Postorino et al. (2019), an extension of the pushback operation during the "warm-up" segment—and the corresponding reduction of the engine ignition time for the aircraft—produces an effective reduction of carbon emissions for the whole taxi-out procedure. In other words, the approach by elements allows identifying actions—in this case a different use of tugs for the whole taxi-out procedure—that provide effective results in terms of carbon reduction. In addition to this, the development of aeronautical and engine technology in the next future will allow reducing fuel consumption and, consequently, the emissions generated by aircraft in one or more segments. For a given reference period and a given airport layout, the carbon taxi-out emissions at the airport can be conveniently simulated and computed by considering the elementary contribution of each segment—both for the current situation and for possible scenarios.

As for the second example, although the general approach is the same (see Fig. 1), however, the goal here is different. In this case, the activity is still decomposed, but the goal is to understand the contribution of sub-activities, rather than single elements. In other words, for composed activities the first level of analysis requires the identification of the carbon contribution of sub-activities as a whole, in order to identify the most relevant ones. Then some further disaggregation makes it possible to identify elementary segments, as in the first example, starting from the most pollutant sub-activities.

Equation (3) allows obtaining an estimate of the carbon contribution of the given activity by using some average values for the aggregate segments (or sub-activities, depending on the disaggregation levels) that are parts of the examined sub-activity. In this perspective, the "handling" activity has been decomposed in eight sub-activities, each one characterized by one or more further levels of disaggregation (see, e.g., "Luggage/Cargo loader" in Fig. 6). The output provides the carbon contribution of the handling activity, and, as in Eq. (2), the product $(F_{tj,i} \cdot F_{cj,i} \cdot \lambda_{(.)})$ provides the contribution of each sub-activity to the whole carbon output due to handling, thus allowing the identification of the most relevant one in terms of carbon impacts.

To conclude, the EbE approach is general and although the chapter focuses on the airport transport function it may be applied as well to other airport functions such as landside operations, energy production and consumption and other terminal buildings activities.

References

Airbus. (2017). *Global market forecast 2017–2036.*

Airport Cooperative Research Program (ACRP) Report 11. (2009). *Guidebook on preparing greenhouse gas emissions inventories.* Washington, DC: Transportation Research Board.

Airport Council International (ACI). (2018). *Airport carbon accreditation program.* https://www.airportcarbonaccreditation.org/. Last access 5 November 2018.

Airports Council International (ACI). (2009). *Guidance manual: Airport greenhouse gas emissions management.*

Ashok, A., Balakrishnan, H., & Barrett, S. R. (2017). Reducing the air quality and CO_2 climate impacts of taxi and takeoff operations at airports. *Transportation Research Part D: Transport and Environment, 54,* 287–303.

Babikian, R., Lukachko, S. P., & Waitz, I. A. (2002). The historical fuel efficiency characteristics of regional aircraft from technological, operational, and cost perspectives. *Journal of Air Transport Management, 8*(6), 389–400.

Boeing. 2016. *Annual report.* http://investors.boeing.com/investors/financial-reports/default.aspx.

Budd, L., Ison, S., & Budd, T. (2016). Improving the environmental performance of airport surface access in the UK: The role of public transport. *Research in Transportation Economics, 59,* 185–195.

de Santoli, L., Mancini, F., Nastasi, B., & Piergrossi, V. (2015). Building integrated bioenergy production (BIBP): Economic sustainability analysis of Bari airport CHP (combined heat and power) upgrade fueled with bioenergy from short chain. *Renewable Energy, 81,* 499–508.

Graham, A. (2014). *Managing airports: An international perspective* (4th ed.). London: Routledge (Taylor & Francis).

Graham, W. R., Hall, C. A., & Morales, M. V. (2014). The potential of future aircraft technology for noise and pollutant emissions reduction. *Transport Policy, 34,* 36–51.

Heinrich, M. T., Kelch, F., Magne, P., & Emadi, A. (2014). Investigation of regenerative braking on the energy consumption of an electric taxiing system for a single aisle midsize aircraft. In *Industrial Electronics Society, IECON 2014 40th Annual Conference of the IEEE* (pp. 3023–3029).

Huang, H., Chen, L., & Hu, E. (2015). A new model predictive control scheme for energy and cost savings in commercial buildings: An airport terminal building case study. *Building and Environment, 89,* 203–216.

IATA. (2016). *Annual review.* http://www.iata.org/publications/Pages/annual-review.aspx.

ICAO. (2011). *Airport air quality guidance manual, Doc 9889.* http://www.icao.int/environmental-protection/Documents/Publications/FINAL.Doc%209889.1stEdition.alltext.en.pdf.

ICAO. (2016). *Environmental report—Aviation and climate change*. https://www.icao.int/environmental-protection/Pages/env2016.aspx.

ICAO, Aircraft Engine Emissions Databank (EEDB). (2018). https://www.easa.europa.eu/easa-and-you/environment/icao-aircraft-engine-emissions-databank. Last access 5 November 2018.

ISPRA, Italian greenhouse gas inventory 1990–2010, National inventory report 2012. http://www.isprambiente.gov.it.

Jou, R. C., Hensher, D. A., & Hsu, T. L. (2011). Airport ground access mode choice behavior after the introduction of a new mode: A case study of Taoyuan International Airport in Taiwan. *Transportation Research Part E: Logistics and Transportation Review, 47*(3), 371–381.

Kılkış, B. (2014). Energy consumption and CO_2 emission responsibilities of terminal buildings: A case study for the future Istanbul International Airport. *Energy and Buildings, 76*, 109–118.

Kuhn, K. D. (2013). Ground delay program planning: Delay, equity, and computational complexity. *Transportation Research Part C: Emerging Technologies, 35*, 193–203.

Liebert, V., & Niemeier, H.-M. (2010). Benchmarking of airports—A critical assessment. In *Proceeding: 17th World Conference on Transport Research*. Lisbon.

Lu, C., & Morrell, P. (2006). Determination and applications of environmental costs at different sized airports—Aircraft noise and engine emissions. *Transportation, 33*, 45–61.

Morrinson, W. (2009). Understanding the complexities and challenges of airport performance benchmarking. *Journal of Airport Management, 2*, 145–158.

Pels, E., Nijkamp, P., & Rietveld, P. (2003). Access to and competition between airports: A case study for the San Francisco Bay Area. *Transportation Research A, 37*, 71–83.

Postorino, M. N., & Mantecchini, L. (2014). A transport carbon footprint methodology to assess airport carbon emissions. *Journal of Air Transport Management, 37*, 76–86.

Postorino, M. N., Mantecchini, L., & Gualandi, E. (2016). Integration between aircraft and handling vehicles during taxiing procedures to improve airport sustainability. *International Journal of Transport Development and Integration, 1*(1), 28–42.

Postorino, M. N., Mantecchini, L., & Paganelli, F. (2019) Improving taxi-out operations at city airports to reduce CO_2 emissions. *Transport Policy, 80*, 167–176.

Ryley, T., Jaafar, E. J., Budd, T., Miyoshi, C., Keith, M. K., Moxon, R., et al. (2013). Sustainable development and airport surface access: The role of technological innovation and behavioral change. *Sustainability, 5*, 1617–1631.

Suzuki, Y. (2007). Modelling and testing the "two-step" decision process of travellers in airport and airline choices. *Transportation Research E, 43,* 1–20.

Train, K. (2003). *Discrete choice methods with simulation.* Cambridge, MA: Cambridge University Press.

Yu, K. N., Cheung, Y. P., Cheung, T., & Henry, R. C. (2004). Identifying the impact of large urban airports on local air quality by nonparametric regression. *Atmospheric Environment, 38,* 4501–4507.

Zou, B., Elke, M., Hansen, M., & Kafle, N. (2014). Evaluating air carrier fuel efficiency in the US airline industry. *Transportation Research Part A: Policy and Practice, 59,* 306–330.

CHAPTER 11

Can Laboratory Experiments Help in Evaluating Emission Trading Schemes? A Pilot Experiment on Aviation Allowances: Lessons to Be Learned

Sabrina Armenio, Angela Stefania Bergantino and Andrea Morone

1 Introduction

Anticipating the market reaction following the introduction of a law, a tax or a new regulation would help decision makers in their ability to predict and, thus, to choose and evaluate the instruments and alternatives available to achieve their specific goals. Understanding the efficiency of the market as a result of one action rather than another, evaluating the

S. Armenio (✉) · A. S. Bergantino · A. Morone
Department of Economics,
Management and Business Law, University of Bari, Bari, Italy
e-mail: sabrina.armenio@uniba.it

A. S. Bergantino
e-mail: angelastefania.bergantino@uniba.it

A. Morone
e-mail: andrea.morone@uniba.it

T. Walker et al. (eds.), *Sustainable Aviation*,
https://doi.org/10.1007/978-3-030-28661-3_11

behavioural attitude of consumers, experimenting with new markets and evaluating a change in the industrial organization would result in greater political effectiveness and speed in achieving the objectives set.

The use of experiments is an excellent way to "predict" reality since through experiments it is possible to study a market and agents' reactions in a controlled environment. This provides, if the experiment is appropriately designed, useful and efficient information for decision makers.

It is not a coincidence that the use of experimental methods went from being almost non-existent in the mid-1960s (Levitt and List 2007) to be used in more than 200 articles in 1998 (Holt 2006). This method is now widely adopted for various social, economic, environmental and other analyses.

The experimental discipline finds different fields of application: from food to waste, from reforestation policies to investment choices, from transport to fiscal policies and measures and so on.[1] However, a very important field where the application of experiments gives fundamental insights is the environment.

The environmental problem is two-sided: on the one hand, we have some citizens who see the environmental threat due, mainly but not only, to health concerns (i.e. the greenhouse effect increases the temperature of the earth, smog caused by traffic increases the problems of asthma, powders created by industries lead to cancer, etc.), and on the other hand, we find the industries, forced to modify or limit or innovate their production system, taking into account the impact that production causes on the environment (particles in the air, water pollutants, waste waters, etc.) Above them, there are governments, national or international political decision makers who decide the maximum emission limits for each productive sector in an attempt to mitigate the incessant growth of environmental damage.

The political decisions in this regard affect both citizens (separate waste collection, limited traffic areas, paying for plastic bags in supermarkets to encourage disuse, etc.) and firms (purchasing polluting permits, emission limits, energy to be used in industrial processes, changes and

[1] For instance, Morone et al. (2018) studied food waste with a natural experiment comparing waste production when food is consumed individually with a situation in which it is consumed in group. In transport, experiments have highlighted the trade-off between parking availability and costs, in terms of time savings and have supported the design of parking policies with respect to tariffs, investments and regulation (Bergantino et al. 2015).

restructuring of the production process or investments in new technology, etc.).

Thus, the criticality of the environment lies in the fact that it is influenced by several actors. It is, thus, essential to anticipate how agents will react to specific requirements, legal impositions and to the technological evolution that aims to reduce environmental externalities. If these have to be adopted by either citizens or firms, it must be as clear as possible to the decision maker what drives agent's decisions into adopting the specific measures. In order to be effective, these have to be accepted and used by consumers and producers: i.e. it could be very expensive to introduce a new "green" method if it is not adopted. To avoid this situation, it is fundamental to use experiments to understand how the propensity of economic agents (consumers or firms), towards technological innovation, can be increased.

However, getting valid results from an experiment is not always easy. The experiment reproduces in a laboratory various scenarios, at times very different from the real world case except for the decision rule which is the object of the analysis. The set of rules on how to set up the experiment and its capacity to represent the specific context of analysis is pivotal for the success of the study. This will be demonstrated in the following of the chapter, with specific examples related to an analysis of the emission trading scheme (ETS), an instrument which has regulated the aviation market for years in Europe.

In this paper, we show a portion of those rules that should be followed during the execution of an experiment. In Sect. 2, we emphasize the importance of environmental analysis and its experimental methods of study. In Sect. 3, we examine the importance of conducting a good experiment by specifying the relevant rules for us to follow. In Sect. 4, we compare a "pilot experiment" (hereinafter pilot) and the related "incentivised experiment" (hereinafter experiment), in Sect. 5, we examine the different results obtained. Section 6 contains some concluding comments.

2 The Environment and Experimental Methods of Analysis

2.1 *Methodologies and Some Cases of Analysis*

The environment is a so-called non-market good. It is widely known and understandable that let alone, the market will not determine a price for it. Various methods for the definition of its value have been proposed in the years in the literature. One of the most widely adopted is the contingent valuation method (CVM). This can be defined as an economic estimation method.

This method had its main diffusion after the 1980s, following the Exxon Valdez oil spill in Prince William Sound (in 1989), a big environmental disaster, where quantitative assessment surveys were used for the first time to quantify damages. From this event, several studies have been carried out on the CVM used and on improving its effectiveness. Following the widespread criticism on the accuracy and reliability of this method, in 1993 the National Oceanic and Atmospheric Observatory (NOAA) called for expert economists to prove the validity of the method and to tackle the criticism in regard to the contingent valuation surveys (Arrow et al. 1993). After the publication of this report, several studies have been carried out on the CVM used and its effectiveness improved (Carson et al. 2003). It was then recognized as a valid method with several limitations. W. Michael Hanemann (1994) states that, compared to the evaluation of natural damage, CVM allows us to measure the values attributed and to demonstrate the compatibility between contingency assessment and economic theory. Nowadays this method is more widely used due to the greater attention that the environment requires.

The CVM can be applied in different ways. The Stated Preference (SP) methods consist indirect request to individuals on their willingness to pay (WTP) for a good, or more rarely, willingness to accept (WTA) to abandon a good (Hanemann 1999; Bateman et al. 2002). From an environmental point of view, the damage created or avoided by using a sustainable lifestyle can be assessed through the average value that the population attributes, for example to "clean" air. This provides a monetary measure of the return of the environmental actions taken by society. The SP method allows a direct economic evaluation of the good that is being analysed. With this method, it is possible to evaluate the environmental preference of agents before the decision makers take the final decision. For example, the Forestry Commission in the UK before modifying the landscape asked to estimate the benefits of alternative reforestation strategies, calculating the WTP of the inhabitants for different types of trees (Hanley et al. 1998).

In the field of bike transport, Tilahun et al. (2007) used a SP survey to analyse individual preferences for five different cycling environments. They showed to all respondents nine video clip presentations (taken from the perspective of cyclists) that compared two structures at a time. The authors found that users are willing to pay the highest price for designated cycle paths, followed by the absence of parking on the street and taking an off-road bike path. Also Abraham et al. (2004) with SP method studied cyclist preferences to evaluate the better cycling route by asking respondents to classify three alternative routes in order of preference. They found that cyclists prefer off-road cycling and low-traffic residential roads. Another use of SP method was to determine the relative value of factors which influence pedestrian activity and walkability (Kelly et al. 2011). The survey participants stated how important some identified attributes are, using a five-point scale (from "not important" to "extremely important"). The factor that was identified as the worst for pedestrians was the volume of traffic followed by cleaning the pavement. Matyas and Kamargianni (2018) used a Revealed Preference (RP) questionnaire about Mobility as a Service (MaaS)—services based on smartphone technology involving dynamic, real-time journey planning and electronic ticketing and payment—to find people's preferences and WTP for flexibility. Respondents are presented with repeated choices between four hypothetical MaaS plans out of which three are fixed plans and one is a menu option. Differently, Ruto and Garrod (2009) used a choice experiment approach to investigate the role that agri-environment schemes (AES) design can have on encouraging the participation of farmers to these schemes, central to achieving policy objectives, finding that farmers require greater financial incentives to join schemes with longer contracts or that offer less flexibility or higher levels of paperwork.

Another type of application of CVM is through the use of pairwise choice. With this method, two distinct alternatives are offered to the subjects that have to formulate a repeated choice among them. It is a simpler method since the subjects no longer have to think about a "value" of the good/service/alternative. They do not have to formulate a specific amount but "just" rank the alternatives. Based on the relative frequency of the choices, the demand curve of the average individual is obtained. It is easy to rely on this method to analyse, for instance, the willingness to switch to new technologies or to adopt a specific strategy.[2]

[2]For instance, Ewing and Sarigöllü (2000) apply it to the choice between different types of vehicles, according to the type of fuel. From their results, the strategies of intervention are evaluated in their effectiveness and are taken in consideration to design policies for a greater diffusion of green innovations.

The growing pressures from global and specific environmental legislation (i.e. Paris Agreement, EU legislation, Kyoto protocol, etc.) and agreements safeguarding the environment through different measures and targets is attracting growing interest in these methodologies and their application.

2.2 The Environmental Regulatory Measures: The ETS

Since 1992, with the International Convention on Climate Change, industrialized countries have been committed to stabilizing greenhouse gas concentrations to levels that prevent their dangerous interference with the climate system, implementing this commitment with the 1995 Berlin Mandates that led to Kyoto Protocol, signed in 1997 and started in 2005. Following these international agreements, the EU implemented various measures, among which the emission trading scheme (ETS). The ETS was born within the Protocol in March 2000 and was then included in the EU directive 2003/87/EC, entering into force in 2005.

Given the increase in the volume of flights,[3] the International Air Transport Association (IATA 2011) supported the inclusion in the EU ETS system of all emissions of flights arriving or departing from any airport located in the territory of the European Union or in an EEA-EFTA country (Norway, Iceland and Liechtenstein). For European flights, the system entered into force in 2012 with Directive 2008/101/EC amending Directive 2003/87/EC so as to include aviation activities in the scheme for greenhouse gas emission allowance trading within the Community.[4] Given the pressures, EU lawmakers allowed operators to temporarily derogate from the EU's inclusion in the EU of flights to and from non-EEA countries to allow the International Civil Aviation Organization (ICAO) to reach a global agreement to deal with aviation issues (known as "Stop-the-clock" decision).

[3]The number of flights performed globally by the airline industry has been steadily increasing since the early 2000's and is expected to reach 39.4 million in 2019. This figure is over one million higher than the prediction for the previous year and represents an increase of over 50 percent from a decade prior (https://www.statista.com/statistics/564769/airline-industry-number-of-flights/).

[4]For greater details on the EU legislation on ETS please see other chapters in the Book and in particular Bergantino and Loiacono (2019).

In October 2016, the ICAO set new targets for the compensation and reduction programme of CO_2 emissions from international aviation, called CORSIA. This agreement forces airlines to control emissions on all international routes; the first two phases of CORSIA (until 2027) shall be made on a voluntary basis. To ensure a continuous improvement, in agreement with the objectives of the Paris Agreement signed in 2016, a periodic review of the regime is required.

The EU ETS is the world's first and largest emissions' market. Encompassing 31 countries (the 28 EU, plus Iceland, Liechtenstein and Norway), it regulates around 45% of all EU greenhouse gas emissions from over 11,000 energy-intensive plants (energy plants and industrial plants) and from the airlines connecting these countries.[5] The objective of the EU is to reduce the emissions recorded in 2005 by 43% by 2030. Europe is currently close to the 20% reduction in greenhouse gas emissions from energy consumption compared to the levels in 1990, in line with targets included in the energy-climate package that entered into force in June 2009. It is still, instead, a long way away for 2030s goal.[6]

The ETS method encourages the reduction of carbon emissions produced by industries through the firms' mandatory purchase of permits (emission certificates) in an artificial market for each ton of CO_2 emitted. It is based on a cap which is set on the total amount of greenhouse gases that can be emitted by installations covered by the system. The cap is reduced over time so that total emissions fall. Within the cap, companies receive or buy emission allowances which they can trade with one another as needed. They can also buy limited amounts of international credits from emission-saving projects around the world. The limit on the total number of allowances available ensures that they have a value, and thus a market is created. After each year, a company must surrender enough allowances to cover all its emissions, otherwise heavy fines are imposed. If a company reduces its emissions, it can keep the spare allowances to cover its future needs or else sell them to another company that is short of allowances. Trading brings flexibility that ensures emissions

[5] Data from the official website of the European Commission, session: Energy, Climate change, Environment in EU Actions. https://ec.europa.eu/clima/policies/ets_en.

[6] Ibid.

are cut where it costs least to do so. A robust carbon price also promotes investment in clean, low-carbon technologies.[7]

3 Experimental Rules

To analyse the effectiveness of the measures and the strategies described above it is necessary to identify an instrument that can lead to some insights on agent's behaviour, or, in other words, on how the operators involved react to the measures introduced. The CVM, either through SP methodologies or Pairwise Choice mechanisms, can support policymakers greatly in studying the decision makers' propensity towards new technologies, analysing the reason behind their choices and strategies. Experiments allow us to reproduce the EU ETS permission auction in a controlled environment. This makes it possible for us to study not only the specific choice on how much to pollute or to reduce pollution through the use of pollution permits, but also long period strategic decisions. In particular, through the experiment it is possible to study changes in the investment decisions of firms: investigate when firms buy new machinery (new technology) that emits less units of CO_2, rather than buying permits for all the emissions emitted by old machinery. We can analyse the dynamics of the firm with respect to polluting decisions and related investments, identifying what price of the auction makes the transition to the new technology economically convenient. Experiments allow us to study firms' strategies and therefore the evolution of the market involved. In particular, to verify the effectiveness of ETS in influencing operator's behaviour: How they react to the measure. The use of laboratory experiments, in fact, allows to reproduce in a controlled framework the market environment, providing useful information for efficient decision-making.

Behavioural predictions, however, must be valid in reality and the actual behaviour of subjects is often different from the results of the theoretical models. Predicting the market through past markets performance may be distorted by bias or bubbles; laboratory experiments help to eliminate these distortions (Wolfers and Zitzewitz 2004). They are also useful when there is no previous market to predict from.

[7]For greater details on the ETS the reader is referred to: https://ec.europa.eu/clima/policies/ets_en.

Laboratory studies of economic variables may present limits in the generality of their results. For example, there are several factors that can affect the results during the experiment, such as the "moral" choice that influences the "right" choice for the subject in comparison with the choice that the subjects would naturally pick if they weren't being observed (Levitt and List 2007).

To limit the external influences and to obtain an outcome that reflects reality, we must follow a series of measures during the design of the experiment and during its development.

In the literature, there are several papers that provide useful guidelines for conducting laboratory experiments (Friedman and Sunder 1994) and field experiments (Harrison and List 2004).

Here, we retrace the process performed for one of our laboratory experiments, using this as an example to explain a series of rules and suggestions that we consider important to conduct an experiment correctly. Our experiment is based on the new environment emission regulation for aviation (ETS-EAA).

3.1 *Our Experiment*

A regulator for each sector establishes the limited number of permits. Permits must be purchased at the auction. Initially the permits are sold at a low price. The firms calculate the number of permits they want to buy at that price, given their technology and make a quantity request. If the total demand for permits from all firms in the sector is greater than the one decided by the regulator, the price per unit of CO_2 increases, and the auction will reopen with a new price per permit. The auction ends when the total quantity is equal to or less than the one established by the regulator. The permits, however, are not all for sale at the auction. A number of them are freely distributed (grandfathering method) and a percentage, which increases every year, is subject to auction.

In the laboratory, we test the propensity of permits purchase to emit CO_2 by different firms considering the maximum limit established by the regulator. We initially consider a "Myopic" regulator that emits a very high permission limit and subsequently a "Non-myopic" one with a lower number of permits. We also offer the opportunity for firms to change the initial technology or, in other words, to decide whether to "invest" in a technology that emits less units of CO_2 or "not to invest" in it.

We also consider two groups of industries: a group with very similar technologies (homogeneous industry) and a group that has two opposing high and low emission technologies (heterogeneous industry). In total, the experiment involves 4 different combination of treatments (myopic-homogeneous; myopic-heterogeneous; non-myopic-homogeneous industry; non-myopic-heterogeneous industry).

The objectives of the experiment are to (1) verify if firms invest in a new technology or continue their production, i.e. "paying" for the permits and (2) study whether the limit imposed by the regulator influences the choice of firms' investment.

With this experiment, policymakers can evaluate the "ideal" number of authorizations to be issued as the maximum limit in a sector to identify the optimal CO_2 reduction that the market can support.

As observed in the laboratory, the initial assignment of permits—auctioning using an ascending clock auction, grandfathering with reallocation through an ascending clock auction—does not have a significant effect on the choice of adopting new technologies and on the redistribution of permits (Camacho-Cuena et al. 2012).

3.2 *Making the Experiment Understandable*

As a rule, an experiment must be understood by participants. Oftentimes, and especially with regard to software experiments, there are several difficulties to overcome in making the experiment reflects the real market behaviour, such as taxes that only apply based on a choice or auctions that increase based on collective choice. It is essential that participants understand the experiment and procedure. In order to make the experiment understandable, the experimenters have to:

Provide simple and schematic written instructions that are not too long, with examples of choices and results. Sometimes, it may be useful to provide a realistic context (e.g. to introduce our experiment on emissions allowances in aviation, we specified what the new legislation provides for firms, i.e. according to the Kyoto Protocol, firms buy a permit for each unit of CO_2 they produce, explanation of a permit to pollute, the costs and the alternative to buying the permit, etc.).

When using experimental software, even if the subjects understand their choices in the instructions, they may not connect the choices with the screens and the sequence of the experiment. A presentation containing screenshots that anticipate the step-by-step view of what they will do during the experiment destroys the surprise effect and enables participants to focus on the actions to be undertaken.

In case of experiments with double choices or bifurcation of the game, which involves different results or gains based on the choice made, it is advisable to take different practice periods, advising subjects to make a choice first and then inform them after another one of the results that will be obtained. In our experiment, the subject can decide before buying permits whether to buy a new technology less polluting by spending a fixed sum or stay with the given technology. Therefore, we have made two practice periods, spurring the subjects to first make a choice (invest in the new technology) and then the other (not to invest and keep the technology randomly assigned) to realize how the result varies. In this way, the consecutive rounds of play can reflect more conscious choices by reducing the learning effect during the rounds of the experiment.

Finally, it is always useful to incite individuals to clarify doubts by asking questions, support them during the practice periods and ask them to answer to written questions after reading the instructions of the experiment.

3.3 *The Importance of Payment and "Counterweight"*

It is known that a monetary reward is an excellent way to recruit volunteers to participate in an experiment. Furthermore, encouraging them to understand the experiment, yields reasoned choices instead of randomly chosen alternatives, as the participants know they can gain more at the end of the experiment. Linking participants' decisions in the experiment to a monetary payoff leads participants to be more accurate in their responses and allow experimenters to get better results.

For all experiments, and especially for environmental ones, an important element for reflecting reality is the creation of the right "counterweight" among the choices that the subjects make.

Let's say that subjects have to choose between a non-polluting product that involves little gain and a pollutant that results in a greater gain. In real life, they could pay more attention to the environment, but in the experiment their choices might be affected by the fact that they are not "really" buying a harmful product; they simply choose what makes them earn more in that scenario.

It is therefore essential to find the way to create a "counterweight" to the choice. In their experiment about choosing a completely recyclable table and a non-recyclable one, Camacho-Cuena et al. (2004) found that the right counterweight was providing two real tables as a random pay for their subjects. This encourages participants to really choose what they prefer.

Other methods include creating ecological funds that are activated if a part of the participants in the experiment makes the ecological choice over the other. The fund represents the lowest smog/waste/pollution that is given, and the choice obviously represents the WTP to obtain these values.

In some experiments, a good counterweight method is to give a total amount of money (expressed in Experimental Currency Units [ECU]) from which the amounts of the various choices must be removed. This method is useful when the subjects must identify themselves with the role of firms in the market because it encourages a profit maximization behaviour.

This is exactly what we used in our experiment. We gave an endowment of 1600 initial ECUs to each participant and they were first asked to choose whether to invest in the new technology (cost 580 ECU) and then to by the permits to pollute, starting from a price of 5 ECU. The initial sum, 1600 ECU, was not chosen accidentally. We have considered the duration of the one-hour experiment and considered the average hourly wage in Italy (6/7 €). The amount was defined consequentially in order to compensate for the engagement.

From the moment, the game is understood, these results are positive because they reflect the reality of the market and the true intentions of firm strategy.

3.4 *Quantity of Data Collection for Each Experimental Type*

It is quite important for the validity of the experiment to have a certain number of observations because a bigger sample reflects the "average value" of reality. Since the observation number has no fixed measure, it depends on the type of data collected in an experiment.

You can collect data in different ways:

Questionnaire: a series of paper questionnaires are distributed within a classroom, a laboratory or a room, which the subjects must complete. These questionnaires are called "self-compiled" because they do not require the intervention of the interviewer; the questions and explanations are written. For this reason, we must follow a series of rules to simplify and guide the compilation: it must be short, concise and as simple as possible to understand by the subjects (Corbotta 2014). Although there is an experimenter who distributes the questionnaires and assists in the compilation and withdrawals, there are many risks in using this method, such as the potential lack of answers, the lack of control on the compilation (if it is filled with interest or superficially), and the limitation on the number and complexity of the questions. To limit these drawbacks, it is necessary to have several questionnaires proposed so that the missed answer or the answer given with disinterest is compensated by many valid answers. A subject's full attention to the compilation can be obtained through a remuneration, but it is generally difficult to calculate the appropriate payment for the participants. Filling a questionnaire lasts a few minutes, and the participants must be numerous; therefore, if the payment is commensurate with the time taken and all participants have to be paid, the remunerative amount could be too low and to create that extra attention. Furthermore, the questionnaires consist often of a mere compilation of questions. A remuneration would be the same whether the answer given by the subject is reasoned or not. This lack of "counterweight," as a requirement described above, makes payment on this type of data collection less effective.

Laboratory experiment: complex experiments are generally carried out when using software, which requires not only that the subjects answer questions but also they perform interactions. Software in an auction, for example, can instantly indicate to the subjects the current amount of the auction and its increase in a timely manner. However, in addition to this type of impact studies, laboratory experiments can also be based on the judgement that the subjects attribute to certain choices (judgement studies) and on the observation of the subjects (observational studies), as classified by Aronson et al. (1985).

The laboratory environment is much more controlled than the other data collection methods; therefore, there is no risk of having unanswered questions. Generally, laboratories contain a limited number of subjects, but the arrangement of computers is such that communication between them is avoided, allowing subjects to concentrate on the task to perform. Generally, the subjects repeat the same game for many turns and periods to collect more data and possibly "learn" from previous mistakes. In addition to having data from playing periods, it is also good to "change" the game team. This allows the experimenters to have new "pure" observations or new subjects who are no longer "experts" and who play the experiment for the first time. Thanks to the series of controls that the laboratory environment allows, the number of observations of an experiment can be reduced, compared to the number of questionnaires. Furthermore, it is easier to establish the remuneration of the subjects, either immediately or subsequently, by exclusively collecting the generality of the subjects and receiving the remuneration thereafter.

Online data: the use of the Internet to carry out experiments is increasingly widespread. Thanks to the Internet, it is much easier to recruit "volunteers" for participation. We need to differentiate two types of use:

- The Internet is useful for administering questionnaires (always easier due to free and simple applications, such as those offered by Google forms). These online questionnaires have the same disadvantage of the paper ones (questions and length of the questionnaire must be short and concise; possibility of incompleteness and distraction in the compilation). Yet, thanks to the online diffusion, it is possible to have a much higher number of participants in a very short time. In addition, the participants choose when they participate in the survey, thus establishing greater availability in the compilation.
- The Internet uses real platforms and can replicate the experiment in the laboratory. There are more and more crowdsourcing Internet services that allow the requester to coordinate external resources to perform a specific project. This option creates an online job market as employers search for potential workers through easy forms of online payment. Many researchers therefore seek online participation in experiments, paying the participating subjects as if the experiment took place in the laboratory and according to the outcome and choices established during the game. Several authors have shown how the results obtained through the online labour market are perfectly in line with the results obtained in a laboratory, whether they are comparisons of demographic analysis (Rand 2012), or classic experimental designs, like the prisoner's dilemma (Horton et al. 2011). In any case, these comparisons were made with more than double the observations and one-tenth of the online payment, compared to the one in the physical laboratory.

3.5 Pilot Experiment

An experiment in the laboratory requires a series of rules that are impartial. The participants need to be motivated with a fair remuneration. In addition, the experimenters must have the instructions ready and worded properly to guarantee that the subjects understand the experiment. It should all be "calibrated" in words that do not influence the choices of the players. For example, saying that a firm produces more pollution can create a negative connotation. Instead, the word "emission unit" should be used as it implies a necessary process production. When designing the experiment, the first time, there may be unforeseen and uncalculated errors: there may be an oversight in the instructions after numerous revisions, and the experimenter may have difficulty in calibrating the words to explain the game in the presentation (i.e. a screen with a favourable outcome for the choice "Invest" and an unfavourable choice "Do not Invest" could influence the players). In addition, the software itself may have oversights: as much as the experimenter tries in the design with a large number of PCs, there is no actual interaction and therefore small errors may arise (for instance when selecting the choice for the auction or when calculating the final score or among other critical moments). Given the budget spent and all the work behind the experiment preparation, it would be unfortunate for the experiment software to malfunction, as it would compromise the result of the experiment. A pilot would allow the experimenters to test everything and make sure it functions, without having to repeat the experiment, especially when that means recruiting new subjects who have not already done the experiment.

4 The Experimental Design

The pilot and the experiment were conducted in the ESSE Laboratory of the University of Bari using Z-Tree experimental software (Fischbacher 2007). The main difference between the pilot and the experiment is that in the pilot participants, volunteering economics students were not paid for their choices and in the experiment they did receive a monetary payoff correlated on their choices in the experiment. We implemented a 2×2 experimental design where the treatment variables were:

the technological heterogeneity of the firms in the industry, i.e. subjects have different production technologies as to the level of emissions produced for a given quantity of output and

Table 1 Experimental design

	Homogeneous industry	*Heterogeneous industry*
Myopic regulator	T1	T2
Non-myopic regulator	T3	T4

the level of emission permits issued and auctioned off, as the regulator acts myopically or not when deciding the emission cap to be applied.

Table 1 illustrates the four treatments implemented (T1–T4). As it can be seen, there are four combinations of the two treatment variables: homogeneous and heterogeneous industry (as to technology levels) coupled with myopic and non-myopic regulator.

In the pilot, we run one session, with 8 subjects, for each treatment. In the experiment, we run four sessions, with 8 subjects, for each treatment. In total, 160 subjects participated in our study, 32 in the pilot and 128 in the (incentivized) experiment.

As soon as participants arrived in the lab, they were randomly assigned a seat. After reading the instructions, the experimental setting was explained out loud, following the best practices in experimental design and implementation. Before starting the session, two practice periods were played to familiarize the subjects with the software. In the experiment before the practice periods, the understanding of the instructions was tested using control questions.

We used eight different conventional technologies (CT) from 1 to 8, see Table 2 which displays the default emission as well as the abatement cost function for the different technologies. Technologies differ in the default emissions (emax) if no abatement effort is made by the firms. Note that the first one, CT1, denotes the technology with the highest default emissions (i.e. 20), while the last one, CT8, denotes the technology with the lowest default emissions (i.e. 13). In all treatments, firms have the opportunity to adopt a new technology (NT) that significantly reduces the default emissions of any conventional technology. Adopting the new technology implies that firms should invest 580 ECU.

The firm with the first technology, CT1, must have a permit for each emission of pollutants or abate the remaining units without permission

Table 2 Technologies, emission and abatement cost function

Emissions									*Abated emissions*		
CT1	*CT2*	*CT3*	*CT4*	*CT5*	*CT6*	*CT7*	*CT8*	*NT*	*Marginal cost*	*Total cost*	*Abated units*
20	19	18	17	16	15	14	13	7	0	0	0
19	18	17	16	15	14	13	12	6	10	10	1
18	17	16	15	14	13	12	11	5	20	30	2
17	16	15	14	13	12	11	10	4	30	60	3
16	15	14	13	12	11	10	9	3	40	100	4
15	14	13	12	11	10	9	8	2	50	150	5
14	13	12	11	10	9	8	7	1	60	210	6
13	12	11	10	9	8	7	6	0	70	280	7
12	11	10	9	8	7	6	5		80	360	8
11	10	9	8	7	6	5	4		90	450	9
10	9	8	7	6	5	4	3		100	550	10
9	8	7	6	5	4	3	2		110	660	11
8	7	6	5	4	3	2	1		120	780	12
7	6	5	4	3	2	1	0		130	910	13
6	5	4	3	2	1	0			140	1050	14
5	4	3	2	1	0				150	1200	15
4	3	2	1	0					160	1360	16
3	2	1	0						170	1530	17
2	1	0							180	1710	18
1	0								190	1900	19
0									200	2100	20

(e.g.: 20 permits, or 19 permits and abate one unit, or 18 permits and abate two units, etc.). The abatement cost is known and increases for each additional unit.

The auction starts with a price of 5 ECU. The firm with the first technology has the advantage to buy 20 permits because the cost of abatement of a unit is greater (10 ECU). If the requested permits exceed the limit established by the regulator, the auction does not end and reopens at a price of 15 ECU. The firm has the convenience to buy 19 permits at a price of 15 ECU and abate a unit at 10 ECU. If the auction continues to grow, the units that should be abated will always be greater. For example, at auction price of 55 ECU, the firm that wants to maximize its profit can abate the first unit (cost 10), the second (cost 20) up to the fifth unit (cost 50 ECU, a total cost of 5 abatement units is 150 ECU), as these five units would cost less than buying each permit at 55 ECU.

The abatement cost from the sixth unit onwards is 65 ECU, which is higher than the corresponding permit price in the auction. Therefore, it is profitable for the firm with an auction price of 55 ECU for the first technology firm to buy 15 permits and abate 5 units.

We have avoided providing negative connotations by avoiding the expression "pollution" (using the term "unit emitted" instead), as well as avoided positive connotations about the eco-compatibility of new technology or reducing the number of emissions.

In the homogeneous treatments, all 8 technologies are distributed among firms, i.e. one firm receives CT1, one firm receives CT2 and so on till the last firm that receives CT8. In heterogeneous treatments, four firms receive CT1 and four firms receive CT8.

As second treatment variable, we consider the myopic behaviour of the regulator when setting the emission cap to the regulated industry. The emission cap that maximizes social welfare is that for which the industry aggregated marginal abatement cost equals marginal damage. We assume that the regulator acts myopically if they do not consider the potential dynamic efficiency of the emission control policy, that is, the fact that firms might adopt cleaner technologies (NT in our experimental setting). In this case, the regulator considers only the information on conventional technologies to decide the maximum level of emissions that is socially optimal. However, the regulator would act as non-myopic if she incorporates the potential technological change provoked by the emission control policy when setting the industry emission cap.

Table 3 displays the optimal emission cap and the corresponding equilibrium permit price for each one of the implemented treatments.

Table 3 Optimal emission cap and equilibrium price for each one of treatments

	Myopic regulator	*Non-myopic regulator*	*Myopic regulator*	*Non-myopic regulator*
Emission cap (E*)	92	62	92	56
Marginal damage	46	31	46	28
AMAC	50	30	50	30
P*	53	35	55	35

*Indicate "the optimal" Emission Cap

As soon as they arrived at the lab, every subject was randomly assigned a seat. After reading the instructions, the experimental setting was explained aloud following the best practices in experimental design and implementation. Before starting the session, two practice periods were played to familiarize the subjects with the software. In the experiment before the practice periods, the subjects understanding of the instructions was tested through some control questions.

The auction starts with a price of 5 ECU. The firm with the first technology has the advantage to buy 20 permits because the cost of abatement of a unit is greater (10 ECU). If the requested permits exceed the limit established by the regulator, the auction does not end and reopens at a price of 15 ECU. The firm has the convenience to buy 19 permits at a price of 15 ECU and abate a unit at 10 ECU. If the auction continues to grow, the units that should be abated will always be greater. For example, at auction price of 55 ECU, the firm that wants to maximize its profit can abate the first unit (cost 10), the second (cost 20) up to the fifth unit (cost 50 ECU, for a total cost of 5 abatement units is 150 ECU), as these five units would cost them less than buying each permit at 55 ECU per permit. The abatement cost from the sixth unit onwards is 65 ECU, which is higher than the corresponding permit price in the auction. Therefore, it is profitable for the firm with an auction price of 55 ECU for the first technology firm is to buy 15 permits and abate 5 units.

We have always avoided providing negative connotations by avoiding the expression "pollution" (using the terms as a "unit emitted" instead and not polluting), as well as avoiding positive connotations about the eco-compatibility of new technology or reducing the number of emissions.

We therefore divided, both the pilot and the experiment into two parts. In the first part, a myopic regulator issued 92 permits each period. Subject decided n periods under these rules. In the second part, a non-myopic 56 (62) permits were issued per period in the heterogeneous (homogeneous) treatment. Subjects decided then n periods under this new setting.

Each period was independent and consisted in of the following stages:

1st stage: every subject is randomly assigned a conventional technology according to the corresponding distribution (see Table 2). At the beginning of each period, a default profit of 1600 ECU is given to each of them.

2nd stage: all subjects have to decide whether to keep their conventional technology or adopt the new one and invest 580 ECU.

3rd stage: emission permit auction: the emission permits issued by the regulator are auctioned off among the firms. Each subject is allowed to participate in the emission auction. The total number of emission permits released by the regulator (emission cap) is known to all subjects. Subjects are not allowed to buy more permits than their maximum (default) emission level. We use an ascending clock auction.[8] The auction starts at a price of 5 ECU per permit. Then firms are asked to introduce simultaneously their demand for permits at this price. If the aggregated demand exceeds the emission cap set by the regulator, the permit price increases by 10 ECU and the firms are asked to introduce the new demand. If the aggregated demand for permits is equal or lower than the emission cap set by the regulator, the auction ends and permits are distributed according to the firm demand.[9]

At the end of each round, the participants will see a summary screen showing the investment costs (if they have invested in the new technology), the number of units that have been abated and the total cost of units, additionally they can also see the number of permits purchased and the final price of the auction.

4th stage: profits are computed considering that firms' emissions cannot exceed the permits they possess at the end of the auction. Firms should therefore abate the difference between their default emissions (emax) and the number of permits purchased in the auction, paying the corresponding abatement cost.

At the end of each period, a summary of all periods is displayed on the computer screen.

At the end of the experiment, the sum of the n periods of one treatment (myopic-non-myopic) is randomly chosen and converted into

[8]Each bidder has a randomly determined private value for one unit of a good. The auction price ticks up at regular intervals, and a bidder can drop out at any time. The auction ends when the number of remaining bidders equals the number of items. Each of these remaining bidders receives one item, and pays the ending price.

[9]We implemented a price increase by 10 ECUs in the auction to discourage any situation where subject was indifferent between purchasing an additional permit or not.

euros at an exchange rate of 1000 ECU: €1 to determine the subject final payoff. This payoff was paid for real in the experiment and was only announced in the pilot.

5 The Results

The results obtained from the pilot and the experiment are quite different. In Table 4, we provide a comparison of the data collection for the pilot and experiment.

In the pilot, the subjects who invested in the new technology, which allowed them to have a lower number of emissions and therefore allowed to buy less CO_2 permits sold at the auction, were very few. On average, 1.7 firms out of 8 invested in a period of Myopic treatment and 3.3 in Non-Myopic treatment, resulting into a large behavioural difference under the point of view of regulation (under a non-myopic regulator, the percentage of investment is almost double compared to Myopic regulation).

In the final experiment, the investment behaviour is clearly different. First, the percentage of agents investing in new technologies is much higher in both treatments. On average the firms that have invested are 3.7 out of 8, namely about half of the firms. Second, there is no difference (if not minimal) between myopic and non-myopic treatment. In Table 5, data on the experiments are compared.

Table 4 Descriptive data collection of the pilot and the experiment

	Pilot	*Experiment*
Subject number for session	16	16
Number of sessions	1	4
Total subjects periods under myopic regulator for session	16	64
	7	8
Periods under non-myopic regulator for session	7	8
Total of observations	224	1024
Retribution of myopic (average)	0 €	5.20 € (max 12.1 €, min 0.75 €)
Retribution of non-myopic (average)	0 €	4 € (max 8.9 €, min 0.55 €)

Table 5 Descriptive percentage and average of investment in pilot and final experiment

	Pilot experiment	*Final experiment*
Myopic treatment		
Percentage of total investment	22.3	45.5
Average firms invest in a period	1.7	3.65
Non-myopic treatment		
Percentage of total investment	41.1	46.7
Average firms invest in a period	3.3	3.75

Table 6 Probit estimation: probability of investing in new technology in the last periods compared to the first ones in the pilot and in the experiment

	Pilot	*Marginal effect*	*Experiment*	*Marginal effect*
Last periods	0.376**	0.128**	−0.0984	−0.03906
	(0.180)	(0.061)	(0.0785)	(0.03113)
Constant	−0.750***		−0.0490	
	(0.123)		(0.0554)	
Observations	224	224	1024	1024

Standard errors in parentheses
***$p<0.01$, **$p<0.05$, *$p<0.1$

It is important to underline that the different results derive from a different understanding of the experiment, which has been improved in the final experiment both by the greater attention of the subjects (knowing that their gain was proportional to the individual result obtained during the experiment) and the explanation and instructions provided in the final experiment.

These explanations and instructions were different than those from the pilot because the questions and clarifications asked in the pilot were inserted or modified in the explanations for the experiment.

To demonstrate how this procedure has influenced the results, we show below the learning effect of the subjects during the various periods of play, analysing whether the probability of investing depended on the period of play and if firms learned during the succession of periods.

We propose a probit analysis on the whole experiment. We have used the 224 observations generated by the pilot and the 1024 by the experiment.

Table 7 One-way ANOVA estimation about difference in mean of profit between treatments in pilot and final experiment

Variables	*Average profit*	*Observation*
Pilot/myopic	4370*** (227.2)	112
Final/myopic	5246*** (106.3)	512
Pilot/non-myopic	3044*** (8227.2)	112
Final/non-myopic	3998*** (106.3)	512

Standard errors in parentheses
***$p<0.01$, **$p<0.05$, *$p<0.1$

Considering the low number of observations of the pilot experiment, we group the periods in 2 sections: first periods (4 in the pilot experiment and 4 in the final experiment) and last periods (3 in the pilot experiment and 4 in the final experiment) instead of segmenting for each period.[10]

Despite the unbalance of the observations in the pilot (7 periods were played in the pilot and thus the division of the periods in two sessions negatively affects the last periods), we can see a significant probability of greater investment in the last periods compared to the firsts. Reading the marginal effect, we can state that in the last periods the probability of investing in new technology is 13% higher than in the first 4 periods ($P=0.037$, $Z=2.08$). Results are reported in Table 6.

As described earlier, the experiment involves 4 different treatments (myopic/non-myopic, homogeneous industry/heterogeneous industry)

When there are behaviour variations during the game periods (e.g. the subjects' greater tendency to invest in the new technology compared to the first periods of the game), we have the learning effect of the subjects: they understand the strategy to obtain a better result.

In Table 6, we report how the investment decision in the pilot and final experiment varies. In the final experiment, we find that there is no

[10]We conducted a One-way ANOVA with the data of the pilot experiment to determine if there was a difference in the grouping of the odd period between the first periods (First periods = 4, Last periods = 3) or between the last periods (First periods = 3 Last periods = 4). The result determined by one-way ANOVA (F (3, 12) = 1.47, p = 0.2711) shows that there was no difference between the two groupings and we can avoid the segmenting for each period.

Table 8 Tukey post hoc test showing the difference in treatments between pilot and final experiment

Comparison	*Final/myopic–pilot/myopic*	*Final/non-myopic–pilot/non-myopic*
Difference (*p*-value)	875.99 (0.003)	954.089 (0.001)

p-value in parentheses

growing trend of investment. In fact, the coefficient of the investment variable is not significantly different from zero, which implies the absence of learning effects in the four treatments. Investments were positive and constant during the experiment.

In the case of the pilot experiment, on the other hand, the probit analysis shows that the probability of investing in new technology is 12.8% higher in the last periods than in the first periods[11]: investments increased during the experiment.

An interesting analysis is to evaluate whether the payment in the final experiment has influenced the player's interest to earn more by making profit maximization. As described in Table 1, in the final experiment, the gain in myopic treatment was on average €4.4, where the maximum remuneration was €12.15 and the minimum €0.7 rounded up to €1 as the minimum wage for participation. In the non-myopic, the average was €4, with a maximum pay of €8.9 and a minimum of €0.5, which also rounded this to €1. In the pilot, there was no payment. However, the subjects had access to a screen with their profit obtained and the cumulative profit of the periods at the end of each period.

A one-way ANOVA (Table 7) was conducted to determine if the average profit was different in the different treatments.

There was a statistically significant difference between groups as determined by one-way ANOVA ($F\ (3, 1244) = 37.52$, $p = 0.000$). A Tukey post hoc test (Table 8) revealed that profit was statistically significantly higher in the final/myopic vs pilot/myopic (875.99 ± 250.8 ECU, $p = 0.003$), and there was statistically higher in final/non-myopic *vs.* pilot non-myopic (954.09 ± 254.09 ECU, $p = 0.001$).

[11]Separating the probit analysis for the 4 treatments and analysing the individual period, we find that the learning effect is maximum in the case of a homogeneous industry under myopic regulation, where the probability of investing in the last period is about 90% higher than the first period. There is no increase in investments during the periods only for a heterogeneous industry under a non-myopic regulator.

The comparative analysis of the experiments' results shows the relevance in defining the rules of the experiment when aiming to use experiments to draw insights into policy interventions. The participation of the respondents improves significantly when they understand the rules and they know that the experiment is incentivized and not hypothetical. Rules are understood better and the behaviour of participants tends to adapt to the rules quicker than when no payment is assigned.

6 Conclusions

In this article, we want to underline the importance and the increasing use of incentive-compatible laboratory experiments that allow policymakers to analyse the progress and results of their actions by simulating market behaviour quickly and with very limited costs.

Experiments are recognized as a valid method even for those sectors where there is no market, namely the environment. Environmental regulations increasingly impose restrictions on the emission limit. We show that while for regulation that introduces the obligation to reduce emissions derived from agriculture, transport, construction and waste (regulation n.2018/842/EU), people propensity for this change can be analysed by pairwise choice (i.e. the choice between an electric and a gasoline fuelled car), Emission trading schemes (Directive 2003/83/EC and following) can be studied in a laboratory context (simulation of the auction of permits).

We stress, however, that to use the result of the experiments as a valid resource for implementation in the real market, it is necessary to follow certain rules that guide the experiment, otherwise biased result would be obtained. Among these rules, we identify the importance that the subjects identify themselves with the context replicated by the experiment: they can be asked to identify themselves in the role of a firm or simply to choose as they would actually choose between two objects or situations. Payment of the subject aids in the identification, especially when it is commensurate with the performance, and also if there is a counterweight in the choices made.

An important rule for us is the use of the pilot. We have shown how, thanks to the pilot, we can test the software and improve the teaching approach so that the identification of the experiment subjects in the real market is clearer.

By comparing the results of the pilot, with the results of the experiment, we have been able to demonstrate how adherence to the experimental rules is fundamental to the results of the experiment.

In the pilot, the payment was not made to the subjects. The use of monetary incentives in the laboratory provides a reliable method to simulate behaviour in the natural environment (Harrison et al. 2007) so the lack of incentives leads to the application of fewer strategies in the choice.

In the incentive-compatible experiment, however, the subjects were paid in proportion to the earnings they have made, so their interest in earning is comparable to the firms reality of maximizing their corporate profits. The results show how, by paying the experiment subjects, the profit maximization is higher in the final experiment than in the pilot. This difference is statistically significant.

We have also analysed the behavioural difference during playing periods: in the pilot, the decision of the subjects facing dichotomic choices grew during the periods, indicating that in the last periods they have learned the strategies to adopt. In the experiment, this strategy was adopted from the beginning because they had a better understanding and identification of the experiment.

The results show how, following the rules indicated in our article, which are:

i. a greater understanding of the experiment by individuals with clear identification of their role in the experiment,
ii. the use of monetary compensation,
iii. the clearer explanation and organization of the compensation, and
iv. the use of a pilot that allows to test the weaknesses of the experiment; experimenters can obtain solid results, reflecting the behaviour of people outside the laboratory environment, helping the decision makers to adopt better solutions.

References

Abraham, J., McMillan, S., Brownlee, A., & Hunt, J. D. (2004). *Investigation of cycling sensitivities.* Presented at 81st Annual Meeting of the Transportation Research Board, Washington, DC.

Aronson, E., Brewer, M., & Carlsmith, J. M. (1985). Experimentation in social psychology. In G. Lindzey (Ed.), *Handbook of social psychology* (Vol. 1, pp. 441–486). Hoboken: Wiley.

Arrow, K., Solow, R., Portney, P. R., Leamer, E. E., Radner, R., & Schuman, H. (1993). Report of the NOAA panel on contingent valuation. *Federal Register, 58,* 4601–4614.

Bateman, I., Carson, R. T., Brett Day, W., Hanemann, M., Hanley, N., TannisHett, M. J.-L., et al. (2002). *Economic valuation with stated preference techniques: A manual*. Northampton, MA: Edward Elgar.

Bergantino, A. S., De Carlo, A., & Morone, A. (2015, April 25). *Individuals' behaviour with respect to parking alternatives: A laboratory experiment*. MPRA Paper No. 63815, posted 13:11 UTC.

Bergantino, A. S. & Loiacono, L. (2019). The impact of the European Union Emission Trading System on the aviation sector. In *Sustainable aviation: The greening of the flight path*. Palgrave, forthcoming.

Camacho-Cuena, E., García-Gallego, A., Georgantzìs, N., & Sabater Grande, G. (2004). An experimental validation of hypothetical WTP for a recyclable product. *Environmental & Resource Economics, 27*, 313–335.

Camacho-Cuena, E., Requate, T., & Waichman, I. (2012). Investment incentives under emission trading: An experimental study. *Environmental Resource Economics, 53*, 229–249.

Carson, R. T., Mitchell, R. C., Hanemann, M., Kopp, R. J., Ruud, P. S. A., & Paul, A. (2003). Contingent valuation and lost passive use: Damages from the Exxon Valdez oil spill. *Environmental & Resource Economics, 25*, 257–286.

Corbetta, P. (2014). *Metodologia e tecniche della ricerca sociale* (2° edizione). Il Mulino – Scienze sociali, ristampa.

Ewing, G., & Sarigöllü, E. (2000). Assessing consumer preferences for clean-fuel vehicles: A discrete choice experiment. *Journal of Public Policy and Marketing 19*(1), 106–118.

Fischbacher, U. (2007). z-Tree: Zurich toolbox for ready-made economic experiments. *Experimental Economics, 10*, 171–178.

Friedman, D., & Sunder, S. (1994). *Experimental methods: A primer for economists*. Cambridge: Cambridge University Press.

Hanemann, W. M. (1994, Fall). Valuing the environment through contingent valuation. *Journal of Economic Perspectives, 8*(4), 19–43.

Hanemann, W. M. (1999). Neo-classical economic theory and contingent valuation. In I. J. Bateman & K. G. Willis (Eds.), *Valuing environmental preferences: Theory and practice of the contingent valuation method in the US, EU, and developing countries*. Oxford: Oxford University Press.

Hanley, N., Wright, R. E., & Ademowicz, V. (1998). Using choice experiments to value the environment. *Environmental & Resource Economics, 11*(3–4), 414–428.

Harrison, G. W., & List, J. (2004, December). Field experiments. *Journal of Economic Literature, XLII*, 1009–1055.

Harrison, G. W., List, J., & Towe, C. (2007, March). Naturally occurring preferences and exogenous laboratory experiments: A case study of risk aversion. *Econometrica, 75*(2), 433–458.

Holt, C. A. (2006). *Markets, games, and strategic behavior: Recipes for interactive learning*. Boston: Addison-Wesley.

Horton, J. J., Rand, D. G., & Zeckhauser, R. J. (2011, September). The online laboratory: Conducting experiments in areal labor market. *Experimental Economics, 14*(3), 399–425.

International Air Transport Association—IATA. (2011). *Vision 2050*. IATA Report.

Kelly, C. E., Tight, M. R., Hodgson, F. C., & Page, M. W. (2011). A comparison of three methods for assessing the walkability of the pedestrian environment. *Journal of Transport Geography, 19*, 1500–1508.

Levitt, S. D., & List, J. A. (2007, Spring). What do laboratory experiments measuring social references reveal about the real world? *Journal of Economic Perspectives, 21*(2), 153–174.

Matyas, M., & Kamargianni, M. (2018). Survey design for exploring demand for Mobility as a Service plans. *Transportation*. https://doi.org/10.1007/s11116-018-9938-8.

Morone, P., Falcone, P. M., Imbert, E., & Morone, A. (2018). Does food sharing lead to food waste reduction? An experimental analysis to assess challenges and opportunities of a new consumption model. *Journal of Cleaner Production, 185*, 749–760.

Rand, D. G. (2012). The promise of Mechanical Turk: How online labor markets can help theorists run behavioral experiments. *Journal of Theoretical Biology, 299*, 172–179.

Ruto E., & Garrod G. (2009). Investigating farmers preferences for the design of agri-environment schemes: A choice experiment approach. *Journal of Environmental Planning and Management, 52*(5), 631–647.

Tilahun, N. Y., Levinson, D. M., & Krizek, K. J. (2006). Trails, lanes, or traffic: valuing bicycle facilities with an adaptive stated preference survey. *Transportation Research Part A, 41*(2007), 287–301.

Wolfers, J., & Zitzewitz, E. (2004, Spring). Prediction markets. *Journal of Economic Perspectives, 1*(2), 107–126.

CHAPTER 12

Milestone of Greening the Flight Path: Alternative Fuels

Selçuk Ekici, İlkay Orhan, T. Hikmet Karakoç and Arif Hepbasli

Energy requirements are increasing day by day for reasons such as population growth, industrialization, and the increasing consumption habits of mankind. In this process, sustainability and the amount of energy used have become one of the most fundamental issues.

The need for energy for economic development is a fundamental requirement. The economic development of countries depends on the sustainability of their energy resources. Energy consumption for increasing population and industry reveals the sensitivity of the use of energy

S. Ekici (✉)
Faculty of Engineering, Department of Mechanical Engineering, Iğdır University, Iğdır, Turkey

İ. Orhan · T. H. Karakoç
Faculty of Aeronautics and Astronautics, Department of Airframe and Powerplant Maintenance, Eskişehir Technical University, Eskisehir, Turkey

A. Hepbasli
Faculty of Engineering, Department of Energy Systems Engineering, Yasar University, İzmir, Turkey

T. Walker et al. (eds.), *Sustainable Aviation*,
https://doi.org/10.1007/978-3-030-28661-3_12

resources. Besides the effects of economic developments, the negative environmental impacts of the increase in the use of fossil fuels should also be considered. In this context, the world is increasingly interested in alternative fuels and renewable fuels (European Aviation Environmental Report 2016).

Alternative fuels have some advantages, such as being more environmentally friendly and coming from more diverse resources than fossil fuels. It is also becoming increasingly important in the socio-economic development of countries. The terms "alternative fuel" and "renewable fuel" are sometimes used interchangeably, although they have specific meanings in laws and regulations. Generally, the term "alternative fuel" is defined as fuel that is used as alternative instead of traditional gasoline and diesel fuel.

The term "renewable fuel" is defined as fuels, such as plants, animal wastes, or municipal solid wastes obtained from renewable sources and non-oil derivatives. Renewable fuels include electricity and liquid–gas fuels derived from renewable biomass energy sources against fossil fuels. Alternative fuels include gas fuels such as hydrogen, natural gas, and propane; alcohols, such as ethanol, methanol, and butanol; and vegetable and waste derived oils (EPA: United States Environmental Protection Agency 2018).

Results of aircraft operations—consuming conventional fuels derived from fossil sources—have caused significant source of adverse emissions. Because of rapidly developing aviation industry, improvement of materials and variation of energy supplement have increasingly been significant. The most important parameters of sustainable aviation are to reduce environmental impacts occurring by the use of energy sources and to increase resource utilization efficiency (Penner 1999).

Turbojet engines, starting from industrial applications, heating, electricity generation, to unmanned air vehicles, military vehicles, missile technology, commercial airplanes, and aircraft auxiliary power units are used in many advantageous areas in our lives. Performance and efficiency of turbojet engines are important both economically and for adapting to advancing technology (Mattingly and Von Ohain 2006).

In this study, renewable and alternative fuels are investigated in the field of aviation. The study has shown that alternative and renewable fuels are desirable for low emissions combustion in gas turbine engines. In addition, the advantages and disadvantages of displacement of fossil fuels with new generation fuels (alternative and renewable) are also mentioned in the article. Alternative and renewable fuels are now being used in all sectors. According to the estimates of scientists, humanity's dependence on fossil fuels will be seriously diminished by the developments in fuel production technology in the next century.

1 Are Biofuels Capable of Replacing Fossil Fuels?

Because of the global warming effect today, all areas are turned to the alternative fuels. The jet fuel, which is obtained from oil is not a renewable fuel. Biofuels and synthetic fuels are two alternative fuels that are offered for the aircrafts. Since the environmental factors push the oil prices to the highest value, aviation industry focuses on the biofuels for the alternative fuels (Blochel 2017).

The first unmanned test flight using biofuel was carried by the cooperation with Boeing, General Electric and Virgin Atlantic Airlines in 2008. The Boeing 747 flew from London to Amsterdam. The used alternative fuel was derived from the coconut and babassu plants. The tank was connected to a single engine of four-engine aircraft. 20% of the fuel was supplied here, resulting in the successful test flight (IATA 2009).

A second test flight was conducted at the end of 2008. During the test flight carried out by Boeing, Rolls Royce, and Air New Zealand Airlines, only 50% biofuels were used in one engine of the four-engine aircraft. Biofuels were obtained from the jatropha plant, and this test flight ended successfully. The third test flight was conducted in cooperation with Boeing, Continental Airlines, General Electric, and Honeywell CFM. This test flight was successfully conducted on January 7, 2009. For this flight, 50% biofuels (algae and jatropha) were used in one engine of the two-engine aircraft. After this test on January 30, 2009, a successful test flight of a four-engine Boeing 747-300 aircraft was carried out. Algae, camel, and jatropha biofuels were added only at 50% for a single motor. This test flight was carried out by Boeing, Japan Airlines (JAL), Pratt & Whitney, and Honeywell (IRENA 2017).

After the test flight, commercial flights began in 2011. Oils used for food production were used in flight. Flight was carried from Amsterdam to Paris by Boeing 737 KLM aircraft. The longest flight with biofuel was made with a Boeing aircraft belonging to Netherlands Airlines. This flight was held on June 20, 2012, and waste cooking oil was used as biofuels. Flights were conducted successfully between Amsterdam and Rio De Janeiro. In all these flights, a certain ratio of biofuels was used.

In 2012, the Canadian National Organizations (NRC) performed a successful flight using entirely biofuel. This flight was performed with Falcon 20, and biofuels derived from carinata oil were used (NRC 2012).

Biodiesel is derived from oilseed plants. Biodiesel is derived from plants such as flax, sunflower, canola, soybean, safflower, cottonseed, and corn. It is a renewable fuel and emerges as crude or refined depending on the processing of vegetable and animal oils. Biodiesel is mixed with diesel fuel used in specific proportions. Diesel/biodiesel mixture ratios are named according to their different forms (Blin et al. 2013). The so-called B2 consists of 2% biodiesel—98% diesel. B5 consists of 5% biodiesel—95% diesel. B20 consists of 20% biodiesel—80% diesel. B50 consists of 50% biodiesel—50% diesel. B100 consists of 100% biodiesel.

Plants used for biodiesel can be converted into oil, sunflower oil, soybean oil, safflower oil, rapeseed oil, peanut oil, flaxseed oil, groundnut oil, cottonseed, and waste frying oils (Hari et al. 2015).

Biodiesels are primary vegetable oil and derived from biological sources with diesel-equivalent processed fuel. Biodiesel, that does not contain petroleum and harm the environment when burned, is a renewable resource that can be produced domestically. Due to their biodegradation properties, biofuels are not suitable for international transport and storage in shelf or tank. Therefore, it is more reasonable to consume within the region where they are produced. Biodiesel does not contain sulfur, so it does not damage the environment. In addition, biodiesel is not toxic and aromatic. The physical properties of biodiesel fuel are similar to diesel fuel. Biodiesel is used in compression-ignition and diesel engines. In some cases, biodiesel can be used with slight changes or without the need for these changes (McDowell Bomani et al. 2009).

Carbon dioxide produced by combustion of biodiesel fuels is converted to oxygen during a process called photosynthesis by plants that used in biodiesel fuel production. As a result, the biodiesel fuel produced from plants and the resulting CO_2 emission in the use of these fuels are balanced by the plants with a cycle. On other hand, the burning of fossil fuels causes the release of CO_2 as an additional burden to nature. When compared to petroleum-based fuels, it releases 60% less CO_2. The reason is that plants produce photosynthesis from atmospheric carbon dioxide.

Biodiesel fuel reduces the sulfate emissions because it comprises less than 15 ppm sulfur. Some biodiesels produce more nitrogen oxides than other biodiesels, and the biodiesel concentration in the fuel increases nitrogen oxide emissions (McDowell Bomani et al. 2009). In research using 100% biofuels, the worst feature in terms of exhaust emissions of biofuels is the increase of 12–15% NOx. Particularly, the NOx increase in soybean oil methyl ester has been observed in maximum amounts. The auxiliary systems, such as EGR and catalytic converters, have been used to reduce NOx emissions for B100 (Ölçüm 2006). Biodiesels have less than 67% hydrocarbon emissions compared to diesel fuel. In addition, when compared to diesel fuel, the carcinogenic effect of biodiesels is less than aromatic hydrocarbons with a high carcinogenic effect (Bünger et al. 1998).

There are some disadvantages to using biodiesel in the aviation industry. Disadvantages to using biodiesel include the freezing point of biodiesel, the cloud point, the case of prolonged storage of biodiesel, and the gelling temperature and thermal stability (Wang and Tao 2016). Biodiesel should be stored away from a warm, clean, dry, and dark environment. The material of the storage tank is also important. Mild steel, stainless steel, fluorinated polyethylene, and fluorinated polypropylene can be used as storage materials. The viscosity of fuel also plays an important role in the ideal operation of the engine. The viscosity affects atomization of fuel, the ignition delay, combustion, and efficiency. As an example of viscosity effect, efficiency decreases due to the high viscosities of vegetable oil esters.

Biodiesel fuels are a more environmentally friendly fuel compared to fossil fuels as they consume less energy during the production process. Use of biodiesel as energy produces 93% more energy than the energy required for biodiesel production. Furthermore, when comparing fossil fuel with biodiesel fuel, biodiesel reduces greenhouse gas emissions by 41%. Biodiesel is the only alternative energy source that has been fully evaluated, and positive opinions have emerged in terms of harmful emissions and health effects within the framework of the EPA program (Bolszo and McDonell 2009; Gupta et al. 2010; Rehman et al. 2011).

Ethanol is a clear, colorless, and nontoxic liquid with a lower thermal value than gasoline. Ethanol is an alcohol obtained by the fermentation of substances, such as sugar, sugar-containing cellulose, or starch, which contain ethyl alcohol. In countries with large agricultural areas, the use of ethanol in engines is widespread.

Ethanol is used with gasoline, mixed with certain proportion. E80 fuel with 20% gasoline mixture has been used as fuel for automobiles in

US agriculture for many years. Because of the large amount of sugar cane in Brazil, ethanol has been used as fuel in cars for more than 25 years.

Bioethanol fuel is derived from renewable sources. Agricultural crops such as grains, corn, sugar beets, sugar cane, potatoes, wood, straw, and household waste are used for the production of bioethanol. The first step in the production of bioethanol is the hydrolysis process, which transforms the plant into sugar. According to the pretreatment carried out for structural and chemical change, cellulose is converted into sugar or glycoside. After this process, sugar is converted to alcohol by the effect of fermentation. After the hydrolysis process, fermentation, distillation, and dehydration processes are carried out. Sugar beet, sweet sorghum, maize, wheat, barley, rice, potatoes, and sugar cane are used in bioethanol production. Bioethanol obtained from sugar cane and corn varieties are widely used. When used alone or mixed with gasoline at different ratios, air pollution and consumption of petroleum products are decreased. Looking at the production of biofuels in 2007, 84% of biofuel production is biodiesel, and 16% is ethanol (Koh and Ghazoul 2008; Bindraban et al. 2009; Phalan 2009; Janaun and Ellis 2010; Solomon 2010; Vermeulen and Cotula 2010; Achten et al. 2010; Mitchell 2011; Harvey and Pilgrim 2011; Janssen and Rutz 2011).

The main topics discussed in the literature on aviation alternative fuels are given below (Contreras et al. 1997; Edwards and Maurice 2001; Maurice et al. 2001; Daggett et al. 2006; Korres et al. 2008; Hileman et al. 2008; Marsh 2008; Nygren et al. 2009; Bulzan et al. 2010; Edwards et al. 2010; Rye et al. 2010; Anderson et al. 2011; Blakey et al. 2011; Hui et al. 2012; Trivedi et al. 2015; Chuck and Donnelly 2014; Zhang et al. 2016; Won et al. 2016; Bann et al. 2017; Kim et al. 2017):

- Why do we need alternative fuels?
- Which of the alternative fuels is suitable for aviation?
- What should be the production methods of alternative fuels?
- What are the combustion and performance relationships?
- What is the best alternative fuel source?
- What type of alternative fuel is used in aviation?
- What research is needed to develop alternative fuels?
- Which country will have the market area?
- What is life-cycle assessment of alternative fuels?

2 Alternative Fuel Test Steps

The testing processes of alternative fuels (AFs) against traditional fuels (TFs) in the gas turbine engines were determined from the literature (Habib et al. 2010; Khalil and Gupta 2013; Mendez et al. 2014; Gupta et al. 2010), as listed in Table 1.

Table 1 Performance analysis criteria

AFs	*TFs*
Thrust	
Thrust specific fuel consumptions	
Thermal efficiency	
Energy/exergy efficiency	
Air compressor outlet temperature	
Turbine inlet and outlet gas temperatures	
Air compressor outlet pressure	
Turbine inlet and outlet gas pressures	
Oxygen and hydrocarbon concentrations of fuels	
Carbon dioxide/monoxide emissions index	
Nitric oxide emission index	

3 Conclusions

One of the most important parameters of comparison of alternative fuels with traditional fuels is thrust. The alternative fuel to be used in the engine is expected to have a close energy content compared to the traditional fuel used. On the sectoral basis, it is required to travel with less fuel consumption for the same distance from alternative fuel, which will be used directly or as a mixture.

Although this is the direction of the theory, unfortunately these values have not been reached in practice. The process results in an increase in fuel consumption of the alternative fuel.

Thrust specific fuel consumption is a parameter that is important in terms of operating cost of the engine and is defined as the amount of fuel consumed to produce propulsion. The consumption of alternative fuel is expected to be low by comparison with cost of AFs. Considering fuel consumption and propulsion, it is desired to be more efficient than traditional fuels. Today, emission indexes have been restricted by many laws. With global warming and air quality problems, especially around the airport, it is necessary to examine the exhaust gases before commercial use of alternative fuels.

Also, in the literature, alternative fuel production method/product (AFs) is evaluated under the following conditions:

- AF's production method
- Thermo-physical properties of AFs
- Shelf life of AFs
- AF's transport
- Easy access to AFs
- Cost of AFs
- Emissions/air quality
- Availability
- Sustainability
- Applicability.

4 Remark

The development of alternative fuels is promising in the solution of today's problems. Although alternative fuels have many problems, we see that alternative fuels are more and more available. We hope that alternative fuels can be substitute for traditional fuels. The disadvantages of alternative fuels can be overcome with technology in the coming years.

References

Achten, W., Maes, W., Aerts, R., Verchot, L., Trabucco, A., Mathijs, E., et al. (2010). Jatropha: From global hype to local opportunity. *Journal of Arid Environments, 74*(1), 164–165.

Anderson, B. E., Beyersdorf, A. J., Hudgins, C. H., Plant, J. V., Thornhill, K. L., Winstead, E. L., et al. (2011). Alternative aviation fuel experiment (AAFEX).

Bann, S. J., Malina, R., Staples, M. D., Suresh, P., Pearlson, M., Tyner, W. E., et al. (2017). The costs of production of alternative jet fuel: A harmonized stochastic assessment. *Bioresource Technology, 227*, 179–187.

Bindraban, P. S., Bulte, E. H., & Conijn, S. G. (2009). Can large-scale biofuels production be sustainable by 2020? *Agricultural Systems, 101*(3), 197–199.

Blakey, S., Rye, L., & Wilson, C. W. (2011). Aviation gas turbine alternative fuels: A review. *Proceedings of the Combustion Institute, 33*(2), 2863–2885.

Blin, J., Brunschwig, C., Chapuis, A., Changotade, O., Sidibe, S. S., Noumi, E. S., et al. (2013). Characteristics of vegetable oils for use as fuel in stationary diesel engines—Towards specifications for a standard in West Africa. *Renewable and Sustainable Energy Reviews, 22*, 580–597.

Blochel, A. (2017). *The future of advanced bio-jet fuel* (Dissertation). Retrieved from http://urn.kb.se/resolve?urn=urn:nbn:se:liu:diva-138629.

Bolszo, C. D., & McDonell, V. G. (2009). Emissions optimization of a biodiesel fired gas turbine. *Proceedings of the Combustion Institute, 32*(2), 2949–2956.

Bulzan, D., Anderson, B., Wey, C., Howard, R., Winstead, E., Beyersdorf, A., et al. (2010, October). Gaseous and particulate emissions results of the NASA alternative aviation fuel experiment (AAFEX). In *ASME Turbo Expo 2010: Power for Land, Sea, and Air* (pp. 1195–1207). American Society of Mechanical Engineers.

Bünger, J., Krahl, J., Franke, H. U., Munack, A., & Hallier, E. (1998). Mutagenic and cytotoxic effects of exhaust particulate matter of biodiesel compared to fossil diesel fuel. *Mutation Research/Genetic Toxicology and Environmental Mutagenesis, 415*(1), 13–23.

Chuck, C. J., & Donnelly, J. (2014). The compatibility of potential bioderived fuels with Jet A-1 aviation kerosene. *Applied Energy, 118*, 83–91.

Contreras, A., Yiğit, S., Özay, K., & Veziroğlu, T. N. (1997). Hydrogen as aviation fuel: A comparison with hydrocarbon fuels. *International Journal of Hydrogen Energy, 22*(10–11), 1053–1060.

Daggett, D., Hendricks, R., & Walther, R. (2006, October). Alternative fuels and their potential impact on aviation. In *25th Congress of the International Council of the Aeronautical Sciences (ICAS 2006)* (NASA/TM-2006-214365, ICAS-2006-5.8.2, E-15568).

Edwards, T., & Maurice, L. Q. (2001). Surrogate mixtures to represent complex aviation and rocket fuels. *Journal of Propulsion and Power, 17*(2), 461–466.

Edwards, T., Moses, C., & Dryer, F. (2010, July). Evaluation of combustion performance of alternative aviation fuels. In *46th AIAA/ASME/SAE/ASEE Joint Propulsion Conference & Exhibit* (p. 7155).

EPA: United States Environmental Protection Agency. http://www.epa.gov/otaq/fuels/alternative-renewablefuels/. Accessed 11 July 2018.

European Aviation Environmental Report. (2016). https://ec.europa.eu/transport/sites/transport/files/european-aviation-environmental-report-2016-72dpi.pdf. Accessed 11 July 2018.

Gupta, K. K., Rehman, A., & Sarviya, R. M. (2010). Bio-fuels for the gas turbine: A review. *Renewable and Sustainable Energy Reviews, 14*(9), 2946–2955.

Habib, Z., Parthasarathy, R., & Gollahalli, S. (2010). Performance and emission characteristics of biofuel in a small-scale gas turbine engine. *Applied Energy, 87*(5), 1701–1709.

Hari, T. K., Yaakob, Z., & Binitha, N. N. (2015). Aviation biofuel from renewable resources: Routes, opportunities and challenges. *Renewable and Sustainable Energy Reviews, 42,* 1234–1244.

Harvey, M., & Pilgrim, S. (2011). The new competition for land: Food, energy, and climate change. *Food Policy, 36,* S40–S51.

Hileman, J., Wong, H. M., Ortiz, D., Brown, N., Maurice, L., & Rumizen, M. (2008, September). The feasibility and potential environmental benefits of alternative fuels for commercial aviation. In *Proceedings of the 26th International Congress of the Aeronautical Sciences* (pp. 5–8).

Hui, X., Kumar, K., Sung, C. J., Edwards, T., & Gardner, D. (2012). Experimental studies on the combustion characteristics of alternative jet fuels. *Fuel, 98,* 176–182.

IATA International Air Transport Association Annual Report. (2009). 65th Annual General Meeting, Kuala Lumpur.

IRENA. (2017). *Biofuels for aviation: Technology brief.* Abu Dhabi: International Renewable Energy Agency.

Janaun, J., & Ellis, N. (2010). Perspectives on biodiesel as a sustainable fuel. *Renewable and Sustainable Energy Reviews, 14*(4), 1312–1320.

Janssen, R., & Rutz, D. D. (2011). Sustainability of biofuels in Latin America: Risks and opportunities. *Energy Policy, 39*(10), 5717–5725.

Khalil, A. E., & Gupta, A. K. (2013). Fuel flexible distributed combustion for efficient and clean gas turbine engines. *Applied Energy, 109,* 267–274.

Kim, D., Martz, J., Abdul-Nour, A., Yu, X., Jansons, M., & Violi, A. (2017). A six-component surrogate for emulating the physical and chemical characteristics of conventional and alternative jet fuels and their blends. *Combustion and Flame, 179,* 86–94.

Koh, L. P., & Ghazoul, J. (2008). Biofuels, biodiversity, and people: Understanding the conflicts and finding opportunities. *Biological Conservation, 141*(10), 2450–2460.

Korres, D. M., Karonis, D., Lois, E., Linck, M. B., & Gupta, A. K. (2008). Aviation fuel JP-5 and biodiesel on a diesel engine. *Fuel, 87*(1), 70–78.

Marsh, G. (2008). Biofuels: Aviation alternative? *Renewable Energy Focus, 9*(4), 48–51.

Mattingly, J. D., & Von Ohain, H. (2006). *Elements of propulsion: Gas turbines and rockets* (pp. 276–277). Reston, VA: American Institute of Aeronautics and Astronautics.

Maurice, L. Q., Lander, H., Edwards, T., & Harrison, W. E., III. (2001). Advanced aviation fuels: A look ahead via a historical perspective. *Fuel, 80*(5), 747–756.

McDowell Bomani, B. M., Bulzan, D. L., Centeno-Gomez, D. I., & Hendricks, R. C. (2009). *Biofuels as an alternative energy source for aviation—A survey.* NASA/TM—2009-215587.

Mendez, C. J., Parthasarathy, R. N., & Gollahalli, S. R. (2014). Performance and emission characteristics of butanol/Jet A blends in a gas turbine engine. *Applied Energy, 118,* 135–140.

Mitchell, D. (2011). *Biofuels in Africa: Opportunities, prospects, and challenges* (pp. 110–111). Washington, DC: The World Bank.

NRC National Research Council Canada. (2012). *Historic milestone for the aviation and sustainable energy industries.* https://www.nrc-cnrc.gc.ca/eng/news/releases/2012/biofuels.html. Accessed 11 July 2018.

Nygren, E., Aleklett, K., & Höök, M. (2009). Aviation fuel and future oil production scenarios. *Energy Policy, 37*(10), 4003–4010.

Ölçüm, T. (2006). *Biodiesel technology.* İstanbul: Yıldız Teknik Üniversitesi Fen Bilimleri Enstitüsü.

Penner, E. J. (1999). *Aviation and the global atmosphere.* Cambridge, UK: Cambridge University Press.

Phalan, B. (2009). The social and environmental impacts of biofuels in Asia: An overview. *Applied Energy, 86,* S21–S29.

Rehman, A., Phalke, D. R., & Pandey, R. (2011). Alternative fuel for gas turbine: Esterified jatropha oil–diesel blend. *Renewable Energy, 36*(10), 2635–2640.

Rye, L., Blakey, S., & Wilson, C. W. (2010). Sustainability of supply or the planet: A review of potential drop-in alternative aviation fuels. *Energy & Environmental Science, 3*(1), 17–27.

Solomon, B. D. (2010). Biofuels and sustainability. *Annals of the New York Academy of Sciences, 1185*(1), 119–134.

Trivedi, P., Olcay, H., Staples, M. D., Withers, M. R., Malina, R., & Barrett, S. R. (2015). Energy return on investment for alternative jet fuels. *Applied Energy, 141,* 167–174.

Vermeulen, S., & Cotula, L. (2010). Over the heads of local people: Consultation, consent, and recompense in large-scale land deals for biofuels projects in Africa. *The Journal of Peasant Studies, 37*(4), 899–916.

Wang, W. C., & Tao, L. (2016). Bio-jet fuel conversion technologies. *Renewable and Sustainable Energy Reviews, 53,* 801–822.

Won, S. H., Veloo, P. S., Dooley, S., Santner, J., Haas, F. M., Ju, Y., et al. (2016). Predicting the global combustion behaviors of petroleum-derived and alternative jet fuels by simple fuel property measurements. *Fuel, 168,* 34–46.

Zhang, C., Hui, X., Lin, Y., & Sung, C. J. (2016). Recent development in studies of alternative jet fuel combustion: Progress, challenges, and opportunities. *Renewable and Sustainable Energy Reviews, 54,* 120–138.

CHAPTER 13

External Benefits of Smart Payment Tools for Airport Train Services

Marco Percoco

1 Introduction

The introduction of advanced forms of payment systems in the public sector is part of a broader context in an attempt to increase the efficiency of public actions by reducing transaction costs and managing operations. Two distinct phenomena contribute to defining the scenario within which the adoption of a new payment instrument is contextualised. On the one hand, there is a substantial afflatus towards the support of urban development through better local public services. On the other hand, a tension arises, certainly due to the contingency of public finances and towards a containment of costs, particularly the operational ones. It is, however, interesting to note that these trends also include an increasing spread of evaluation practices of public actions, both *ex ante* and *ex post* (Percoco 2016).

This paper estimates the environmental external benefits generated by the introduction of a smart payment tool for airport train services. The growing literature on the sustainability of air transport activities and

M. Percoco (✉)
Department of Social and Political Sciences
and GREEN, Università Bocconi, Milan, Italy
e-mail: Marco.percoco@unibocconi.it

T. Walker et al. (eds.), *Sustainable Aviation*,
https://doi.org/10.1007/978-3-030-28661-3_13

of airports in general has mainly focused on the pollution generated by airplanes arriving at or departing from given airports.[1] However, less attention has been paid to how passengers arrive at or depart from the terminals.

In this paper, we argue that the introduction of a smart payment tool based on credit cards may induce passengers to use trains more often and to avoid using other transport modes that are generally more polluting, such as cars and buses.

By making use of estimates from an ad hoc survey for the Malpensa Airport, we argue that the investment costs needed to implement the system are easily counterbalanced by the benefits generated. The results of the sensitivity analysis confirm the social profitability of the investment, even for more extreme situations, suggesting a sort of generalisability of our analysis.

2 The Effects of a New Payment System

2.1 *The Increase in Cost Efficiency and Time Savings*

The introduction of a credit card payment system certainly involves a series of investment costs, but it also provides cost advantages for companies that adopt it. The management of paper tickets implies a series of costs, ranging from the production of the same ticket to the management of the cash and the vending machines.

It is, however, useful to briefly review the cost categories that would undergo a reduction from the introduction of the new payment system.

It is evident that the referenced macro-category must be the one concerning the 'management of the ticket' throughout its life cycle, with financial realisations on five main dimensions:

a. *Costs related to the infrastructure*, namely those related to the purchase, the installation and maintenance of ticket issuers as well as those related to ad hoc validators;
b. *Selling costs*, which are expenses related to the *fees* paid to authorised distributors and the relative management of the cash;

[1] See, for instance, Kaszewski and Sheate (2004) and the essays in Karakoc et al. (2018).

c. *Customer information,* namely the costs related to the ticketing information system;
d. *Revenue protection*, i.e. costs related to the verification of payments to users; and
e. *Production and distribution costs,* i.e. the time spent on the production and distribution of paper tickets.

For the purpose of the present study, reducing ticket management costs has important implications in terms of social welfare. Figure 1 represents the simple case of a transport company that must choose the optimal combination of capital (K) and labour (L) to produce a certain amount of output $Y_{A.}$ For simplicity, it is assumed that if a reduction in ticket management costs involves any change in the cost of labour and capital, the cost frontier moves to the right, reaching a level of production of $Y_B > Y_{A.}$

From an aggregate point of view, this situation is represented in Fig. 2 with a shift to the right of the supply curve, which would entail a price reduction and an increase in the quantity of services consumed on the market. This variation, therefore, implies an increase in social welfare equal to area A.

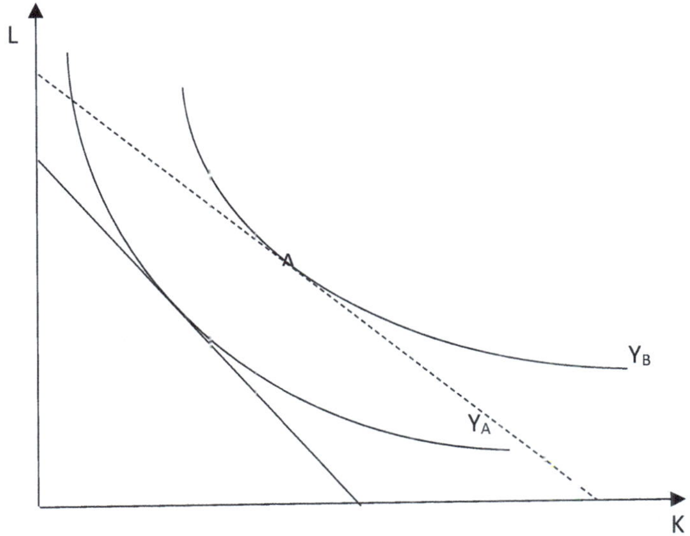

Fig. 1 The effect of a reduction in operating costs

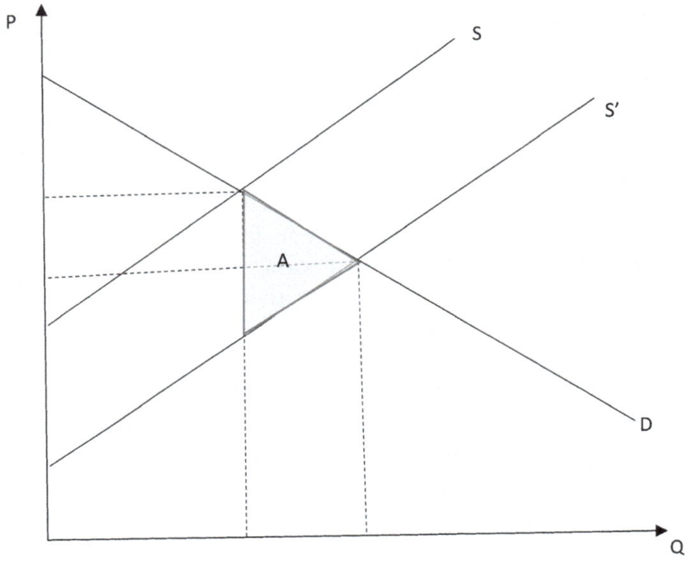

Fig. 2 The effect of a reduction in operating costs on the supply

From an empirical point of view, therefore, the question becomes that of the monetary estimate of area A, which is approximated by the expected variation of operating costs related to the management of the tickets in the case analysed in this study.

The most evident advantage category, and certainly the less problematic to quantify, represents the variation in the time needed to purchase the transport service.

Figure 3, in fact, represents the situation in which the market subject to the introduction of the technological innovation examined here is located. The price variable must be considered, as is usual in the transport sector, to include not only monetary costs (the price of the ticket), but also the time necessary for the use of the service. The use of the credit card involves a reduction of this time as it saves the minutes necessary to purchase the ticket. The price reduction implies, in this case, a change from D to D′, with a consequent variation in social welfare represented by area B. In this paper, we aim to quantify the change in social welfare generated by the introduction of a smart payment tool. Ultimately, the benefits that represent the sum of areas A and B are those that will be quantified here.

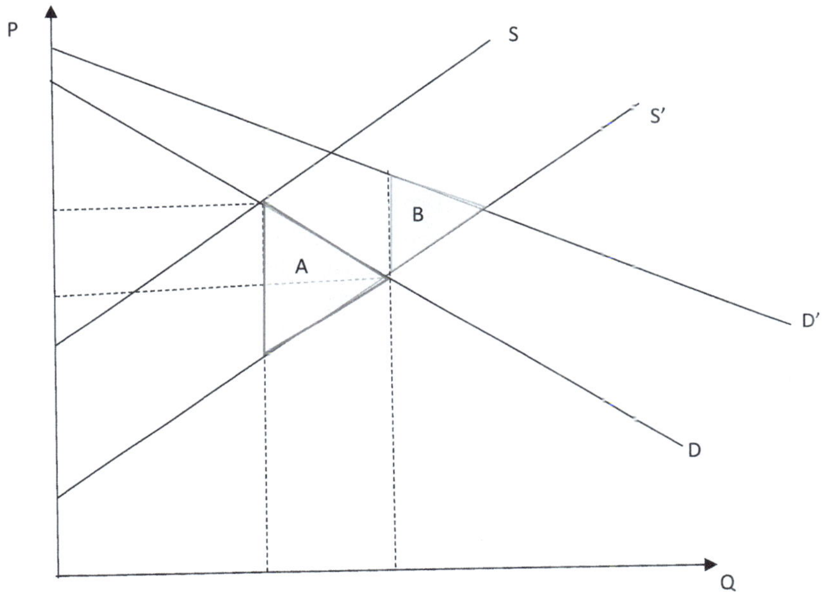

Fig. 3 The combined effect of a reduction in operating costs and travel time

A further comment should be reported and highlighted here. The previous figures discuss the main sources of a potentially positive change in social welfare following the introduction of a new, more efficient payment method. This technological innovation involves a *theoretical* change in the price and quantity of equilibrium *in perfect competition* and, therefore, is not necessarily observable in reality. This theoretical variation, however, is relevant for the purpose of this research report because the cost-benefit analysis (shown in the following sections) refers precisely to conditions of perfect competition, where prices represent the values of goods and resources in general.

2.2 *Additional Non-quantifiable Benefits*

As stated above, the cost-benefit analysis presented in the next section focuses on the benefits related to reductions in costs and time to purchase tickets. Nevertheless, other benefits can be traced back to the introduction of a credit card payment system, although they are not directly quantifiable, especially *ex ante* (Quibria 2008):

a. *Flexibility and convenience.* Using a credit card, consumers have a monetary exit as the service is consumed (or even later, if the transactions are completed at the end of the day). The purchase of paper tickets often consists of an advance payment, even temporally significant.
b. *Security and fraud reduction.* Credit cards, also through an insurance system, guarantee effective protection of the sums available to consumers in case of theft. Similarly, and with regard to transport companies, it is possible to assume a significant reduction in costs deriving from the sale of irregular tickets.
c. *Possibility of price discrimination.* Although this possibility is merely theoretical, given the legislation, the use of the credit card would allow for knowing the consumption habits of users and to infer a willingness to pay individually, and ideally defining a 'personalised tariff'. Although it may appear to be disadvantageous for consumers, this situation of extreme price discrimination would be particularly fair as it would also make significant gains for passengers who are willing to pay below average.
d. *Consumer experience.* As previously stated, the possibility of credit card payment is spreading rapidly in the world's major cities. This phenomenon makes the use of public transport considerably easier and significantly reducing initial information barriers.

All these categories of effects are not easily quantifiable *ex ante*. Nevertheless, the deliberate choice not to make estimates and, therefore, to omit such items within the benefits, is a further reason why the cost-benefit analysis presented in this paper can reasonably be considered not only as a result of prudential estimates, but also as a real lower limit.

3 The Cost-Benefit Analysis Methodology

As stated above, the objective of this essay is to estimate any social welfare gains associated with the introduction of a new urban public transport fare payment system.

However, the technology subjected to assessment envisages the dematerialisation of the ticket as it would be possible to go to the turnstiles or the ticket validator and pay directly for the service. This technology would involve replacing most of the turnstiles and obliterators, which would not only require a new mechanism to read credit cards, but also a network connection to the banking circuit to verify the account user's available funds.

The introduction of a new payment system leads to increased efficiency for operators and time savings for consumers, but with initial investment costs. This characteristic, indeed common to most of the public actions that present characteristics of materiality, led us to adopt the perspective of the cost-benefit analysis for a correct evaluation.

In order to provide an adequate amount of information to correctly read and interpret the results obtained and explained in the following paragraphs, the basic concepts of cost-benefit analysis are summarised below (Bell 2003; European Commission 2014a).

The aim of the economic system is to allocate scarce resources efficiently to maximise the welfare of the community. However, since the level of well-being depends on subjective judgments, it cannot be quantified in absolute terms, nor can interpersonal comparisons be made between changes in well-being. A certain action then follows an increase in well-being if no individual action worsens an individual's condition and at least one improves it (as per the criterion of Pareto improvement).

From a practical point of view, the Pareto criterion is too limiting. The benefits derived from a certain action always involve costs; however, benefits and costs are not distributed among the subjects in the same proportion. We can then define a different situation of potential Pareto improvement, whereby those who derive a welfare gain from a given change can completely compensate those who are damaged. This is a potential improvement only because the level of effective well-being can be influenced by redistributive effects deriving from the event in question when the compensation is not paid.

In other words, while the Pareto improvement criterion postulates a necessarily positive change in well-being for all the stakeholders of a given project, the reality presents interventions that generate flows of costs and benefits that are not necessarily distributed uniformly. The purpose of the cost-benefit analysis is, therefore, to verify that the aggregate benefits are higher than the costs. It should also be noted that the methodology in question is aimed at evaluating only public investments.

In general, cost-benefit analysis is a technique that seeks to bring greater objectivity to the public decision-making process, reducing and highlighting the areas of subjective questionability and discretion, and adopting criteria that allow for compliance with rules of general interest, even when the decisions are taken in a decentralised way. This technique is possible through the identification and quantification in monetary terms of most of the benefits and costs associated with a given project, which is essential for aggregating costs and benefits of a different nature.

Cost-benefit analysis is particularly useful in the field of public decisions when:

- Price signals are inadequate to guide investment decisions;
- Significant spillovers occur, both in terms of benefits and costs, due to the nature and/or size of the project; and
- The well-being of future generations must be taken into consideration.

Since the level of individual well-being depends on a subjective judgment pertaining to the same subject, it cannot be measured in cardinal terms and, consequently, cannot be aggregated to quantify the level of well-being of a group of individuals. One way to circumvent this obstacle is to assume that individual benefits are commensurate with the individuals' willingness to pay in monetary terms (willingness to pay).

If the benefits derived from consumption can be measured by the monetary sum that a person is willing to sacrifice to enjoy the asset rather than renounce it, then the area below the market demand curve measures the aggregate willingness to pay, i.e. a group of individuals. Often, when you do not have an econometrically estimated demand function, the calculation of consumer benefits can be made from the total amount of costs that consumers bear to access a particular asset, which cannot be less than the benefits themselves. For example, in the case of gas, the benefits coincide with the price that the consumer is willing to pay because this coincides, at most, with the perceived utility.

The benefits must be compared to the incurred costs to produce them or to the cost of the factors that have been subtracted from the best alternative employment. These costs do not coincide with those incurred by individuals due to the presence of pecuniary components that are configured as transfers and because of real externalities not 'accounted for' in any private budget. Therefore, they must be corrected and integrated before being included in the balance of the cost-benefit analysis. When the benefits of a given project exceed its costs, the project is economically viable. Benefits and costs are defined as distinct flows that evolve over time according to different rates: in the case of an infrastructure investment, the costs typically exceed the benefits (practically missing) in the construction phase, while in the operating phase, the benefits exceed the costs, and the difference between the two gives rise to a positive net benefit. The comparison between investment costs and net benefits must, therefore, consider the entire life span of the project. A project is economically sustainable if the current net economic value of the complex of these flows is positive:

$$NPV = \sum_{t=1}^{n} \frac{B_t - C_t}{(1+r)^t}$$

where B_t and C_t are the flows of benefits and costs at time t, n is the life horizon of the project and r is the social discount rate.

An alternative indicator of social profitability is the internal rate of return (IRR), which is the discount rate at which the current net economic value is cancelled. As for the financial analysis also in this case, the project has a social advantage if the internal rate of economic return is greater than the social discount rate.

In summary, the following operations must be carried out in the cost-benefit analysis.

1. List all the relevant cost and benefit items and include the different externalities that do not have an immediate monetary manifestation; care must also be taken not to duplicate, i.e. double counts of some items, and not to include pecuniary externality.
2. Evaluate expected benefits and expected costs, first by measuring them and then by giving them a price.
3. Discount the future flow of benefits and costs to determine a capitalised present value; this requires a certain discount rate.
4. Evaluate the project, comparing the aggregated benefits and the aggregated costs. This can be done either by applying the Pareto criterion, i.e. by seeing how much they earn or lose from the different parts involved in the project, or by directly comparing the additional benefits and costs.

It should be noted, however, that the cost-benefit analysis measures the variation of social welfare in a distorted manner; therefore, it correctly assesses public policies only on the condition that the following assumptions are valid:

a. The markets must operate in perfect competition so that the prices represent in an unbiased way the social preferences for the production and consumption of scarce resources;
b. The project to be evaluated must be marginal, i.e. it must not lead to a change in the allocation of resources through a change in relative prices; therefore, it must not have significant indirect effects; and
c. The project to be evaluated must not induce changes in the markets of areas adjacent to its implementation, i.e. it must not have spatial spillover effects through changes in relative prices.

In the case under consideration in this study, assumptions 2 and 3 seem to be easily met, while prices will have to be suitably purified by means of conversion factors so that they can be considered representative of social opportunity costs (European Commission 2014a).

4 The Case of Malpensa Express

4.1 *Project Description*

The project under consideration concerns the introduction of a contactless payment card access system for the Malpensa Express train service between Malpensa Airport (Terminal 1 and Terminal 2) and the centre of Milan (Cadorna, Garibaldi and Centrale stations).

In this way, passengers are offered an 'easy-to-use' service with particular attention given to tourists and non-systematic users who do not know the rules of use/payment for transport and using the 'best do' logic that will calculate, depending on the validations registered in the system, the best rate for the customer for the period considered.

The purpose of the intervention is to offer incoming passengers the possibility of using the railway service in a simpler way, overcoming the intangible barriers constituted by the need to find information with respect to the ticket schemes and how to purchase, taking into consideration that the purchase of the ticket is often difficult for foreigners because of both linguistic difficulties and the lack of foreign currency. The project requires investment in both fixed and mobile equipment as well as in software. For the complete realisation of the project, it is estimated that it is necessary to equip fifty control points, each of which costs €2100. For the company control centre and for the supply of mobile verification terminals, an investment of €100,000 is required.

Overall, the financial expense required to activate the service is €205,000 and is totally concentrated in the first year.[2]

From a qualitative point of view, the social benefits of the project can be classified into the following categories:

a. Effects on the firm adopting the new payment technology as the dematerialisation of the ticket allows for cost savings for the issue and distribution of tickets;
b. Effects on public transit users as they save time by not having to buy tickets in advance, and they understand the pricing mechanism; and
c. If the greater ease of access provides incentives to switch the transport mode, thereby increasing the rail users and reducing the road transport, environmental benefits will be generated

The new payment service is addressed to both current train users and to those who currently use other transport modes but who would be willing to use the train if they were offered greater ease of access. This second segment of demand is of great interest to the railway operator because it determines an increase in revenues without additional costs since the service is not currently saturated. Also, from a socio-economic point of view, this is the segment of demand that produces the greatest benefits because using an unsaturated means of public transportation reduces negative social issues, such as environmental, i.e. producing harmful or climate-altering emissions, and social externalities, such as congestion and accidents.

4.2 *Demand Estimates and Evaluation of Externalities*

To conduct the analysis, it was necessary to reliably estimate how many new travellers would be willing to use the train and which means of transport they were currently using. For this reason, a field survey was developed. This survey was conducted through direct interviews at Milanese airports, which were divided into three phases—February, May

[2]All the data used in the analysis have been provided by Mastercard and SEA, the Malpensa airport operator. SEA is not even in the references! Did the author get the information from the website? Talking to the company? An official document?

and September 2016—for an adequate representation of the various types of customers.

The questionnaire included the following question, which was only addressed to passengers who do not use the train: '*The hypothesis that in the future we can get on the train to and from the airport by paying the ticket in a very simplified way, supporting the credit card at the turnstile, de-materialising the paper ticket, and avoiding having to access a ticket office, is it something that may affect your choice of using the train service?*' The number of passengers moving between Malpensa and Milan city for 2016 was estimated at 5,014,676 throughout the year, equal to 30.6% of passengers '*land side*'. The survey allowed us to estimate that only 27.1% of them use the train. Therefore, 3,657,851 passengers moving between Milan and Malpensa do not use the train.

We must now consider a second important element to assess the number of passengers who could be attracted to the railway with the new payment system: the access system with the card does not allow for multiple purchases of the tickets to avoid the risk of inadvertently double counting. This is a fairly important obstacle for those travelling in groups or with family. To avoid overestimating the number of people who would change their mode of transport, only answers provided by people travelling alone or with colleagues, business partners or friends were selected. In all these cases, it is reasonable to assume that each person pays for their own journey, and in the event of an affirmative answer to the question, that this person is provided with a credit card.

In addition, to try to get an accurate estimate, passengers who travel by car (own, third party or rented) were excluded, considering that in this case, it is difficult for the railway to contend with that mode of transport, albeit with a facilitated access. These criteria, i.e. travel alone and do not use a car, greatly reduce the universe of 'attractive' passengers to 1,241,926, which represents only 34% of those who move between Malpensa and Milan and do not use the train.

These travellers are attracted to the train, 63.7% of whom use the bus services and 36.3% use taxis or rental cars with drivers (NCC).

Table 1 shows the percentage of 'attractive' passengers who responded affirmatively and divided by the means of transport they currently use. Of those travelling between Milan and Malpensa who do not use the train or a private car, and who either travel alone, with colleagues or with friends, 29.1% replied that they would 'certainly' change their choice, while 38% say that they would 'probably' change. Lower,

Table 1 Mode of transport and willingness to use the train with smart payment

Answer *Mode of transport*	*Of course (%)*	*Probably (%)*
A suburban bus line	29.1	38.0
Taxi or NCC	20.4	13.0

although substantial, is the percentage of those using taxis or NCCs who would switch, which is justifiable with the greater comfort offered by these services compared to the train.

We further assumed that among those who answered 'I would certainly change', only 80% would actually do it, while among those who stated 'I would probably change', it was assumed that only 20% would actually do it.

Using these percentages, it was estimated in 2017 that 324,906 passengers would be attracted, which is equal to 8.5% of those who do not use the train. Of these, 78.7% used the regular bus, while the remaining 21.3% used taxis and rental cars with drivers (Table 2).

The average price paid for a ticket is €10.90, with generated revenues amounting to €3,544,430 in the first year, a figure that will grow to €5,078,586 by 2026. In view of these revenues, it should be noted that the ordinary maintenance costs of the equipment are approximately €10,000 per year, and the fee for the credit card transaction is equal to €0.40 per transaction. Total operating costs amounted to €139,962 in the first year, and due to the higher amount of commissions, will increase up to €196,215 by 2026. Table 3 provides a summary of financial cash flows.

By discounting the 10-year cash flow expected at a discount rate of 4% to 2016, a net financial net present value (NPV) of €33,322,069 is obtained.

The expenses incurred for the investment and for ordinary maintenance were converted into economic costs by applying a conversion factor of 0.918. The realisation of the new service will result in the displacement of some passengers from the bus (78.7%) and from the taxi (21.3%) to the train. We are, therefore, faced with a reduction in road journeys without an increase in rail journeys because the rail service has an excess capacity compared to demand and does not require upgrading.

Table 2 Estimates of attracted passengers

	2017	*2018*	*2019*	*2020*	*2025*	*2030*
Total passengers	20,100,000	21,900,000	22,300,000	23,200,000	28,000,000	32,500,000
Transits	2,965,382	3,230,939	3,289,952	3,422,730	4,130,881	4,794,773
Passengers to and from Malpensa	17,134,618	18,669,061	19,010,048	19,777,270	23,869,119	27,705,227
Passengers to and from Milan	5,249,739	5,719,864	5,824,337	6,059,400	7,313,069	8,488,383
Of which they use the train	1,420,426	1,547,628	1,575,895	1,639,497	1,978,703	2,296,709
Of which they do not use the train	3,829,313	4,172,236	4,248,441	4,419,903	5,334,366	6,191,675
Travel alone for business	1,300,141	1,416,571	1,442,445	1,500,660	1,811,141	2,102,218
Attracted passengers	**324,906**	**354,002**	**360,468**	**375,016**	**452,605**	**525,346**

Table 3 Financial cash flows

	2017	2018	2019	2020	2021	2022	2023	2024	2025	2026
Investment	−205,000									
Revenues	3,544,430	3,861,842	3,932,378	4,091,083	4,249,789	4,443,763	4,620,103	4,778,809	4,937,514	5,078,586
Maintenance costs	−10,000	−10,000	−10,000	−10,000	−10,000	−10,000	−10,000	−10,000	−10,000	−10,000
Transaction fees	−129,962	−141,601	−144,187	−150,006	−155,826	−162,938	−169,404	−175,223	−181,042	−186,215
Total	3,199,468	3,710,241	3,778,190	3,931,077	4,083,964	4,270,825	4,440,699	4,593,586	4,746,472	4,882,371

The analysis of the demand led to the aforementioned estimation that 324,906 passengers would be attracted to the railway because of the new payment system in the year 2017. Of these, 21.3% would be subtracted from the use of taxis and rental cars with drivers. Considering a load factor of 1.3 passengers per car, we obtained a reduction of 53,144 trips per year, and this, in turn, would generate a reduction of operating costs that would have to be addressed to produce this service. By calculating the operating costs of private means at €1.02 per kilometre and that the average distance for a trip from and to Milan is 56.35 kilometres, we obtained a reduction in operating costs that would be equal to €3,054,550 in 2017.

The modal shift generates a reduction of 2,994,657 kilometres travelled by taxis in the first year, with a subsequent reduction of external costs in terms of accidents, pollution, greenhouse gas (GHG) emissions, noise and congestion. To evaluate such externalities, we adopted the values reported in the European Commission's (2014b) *Update of the Handbook of External Costs of Transport*, with:

- €178,720 for minor congestion;
- €99,880 for less pollution; and
- €198,610 for minor accidents and other external costs.

Furthermore, by considering a reduction in travel time and a value of time equal to €14.17 per hour, as suggested by the Comune di Milano (2016), we obtained a travel time-savings value equal to €81,578.

Overall, the benefits obtained by reducing road trips were estimated at €3,450,182 for 2017, a value that grows over the years due to the increase in demand to reach €4,943,544 in 2026.

As noted above, the analysis of the demand led to the estimation that 324,906 passengers would be attracted to the railway because of the new payment system in the year 2017. Of these, 78.7% would be subtracted from the bus, which is equal to 255,819 passengers per year. Considering the removal of 20 passengers on every bus run, you get a reduction of 12,791 trips per year, approximately 35 per day. By using an approach similar to the case of the taxis, we obtained the following figures:

- €95,062 for minor congestion;
- €93,688 for less pollution; and
- €102,731 for minor accidents and other external costs.

Using the average parameters relating to the value of the time mentioned above, the value of travel time savings is equal to €1,208,285.

Overall, the movement of part of the traffic from the buses to the train produces social and economic benefits valued at €2,753,279 in 2017, a value that grows to €3,944,997 in 2026.

4.3 Project Appraisal

At this point, we have the necessary elements to carry out a socio-economic analysis of the effects of introducing the new payment system on Malpensa Express at Malpensa Airport and in the stations of Milan North Cadorna, Centrale and Porta Garibaldi.

Table 4 summarises the expected costs and benefits flow over the lifetime of the project from 2017 to 2026.

By using a 3% discount rate, we obtain an NPV equal to €62,431,063. However, this evaluation can be biased by a double margin of uncertainty:

- The field survey is not complete because we waited for the summer period to submit the questionnaire to the type of travellers that characterise this important part of the year; and
- Stated preferences usually diverge from revealed preferences, and this is per se a source of uncertainty.

The European Guide, in cases of high randomness, requires a probabilistic risk analysis: after identifying the critical variables, a reasonable interval must be defined within which the values can be placed. After having established the variability interval and the probability distribution for the critical variables, we proceeded with the calculation of the probability distribution of the project results, in our case, the current net economic value.

As suggested by the European Guide (European Commission 2014a), this operation was carried out by applying the Monte Carlo method.

The variability intervals were defined as follows:

- For those who declared 'I would certainly change', it was assumed that it actually makes a share between a minimum of 75% and a maximum of 90%; and

Table 4 Costs and benefits

	2017	*2018*	*2019*	*2020*	*2021*	*2022*	*2023*	*2024*	*2025*	*2026*
Investment cost	−188,190	0	0	0	0	0	0	0	0	0
Maintenance	−9180	−9180	−9180	−9180	−9180	−9180	−9180	−9180	−9180	−9180
Reduction in operating costs	4,308,063	4,693,860	4,779,593	4,972,491	5,165,389	5,401,154	5,615,486	5,808,384	6,001,282	6,172,747
External benefits	768,691	837,529	852,826	887,245	921,664	963,732	1,001,975	1,036,394	1,070,813	1,101,408
Travel time savings	1,126,707	1,227,606	1,250,028	1,300,477	1,350,927	1,412,587	1,468,642	1,519,092	1,569,542	1,614,386
Net benefits	6,006,091	6,749,815	6,873,267	7,151,034	7,428,801	7,768,293	8,076,923	8,354,690	8,632,457	8,879,361

- For those who declared 'I would probably change', it was assumed that it actually makes a share between a minimum of 25% and a maximum of 50%.

We followed the Guide's recommendations, which warned against the risk of overestimating the positive effects.

The Monte Carlo method requires randomly assigning a value to each variable submitted to the test and subsequently calculating the result obtained with that particular combination of inputs. This procedure must be repeated a sufficient number of times to contain the statistical error within a certain threshold. Thus, 2109 simulations were performed with a combination of two randomly selected variables in the indicated range used for each of them.

The results are reported in Fig. 4. The possible values of the NPV are within the range of €48,044,753 (minimum) and €105,647,692 (maximum). With 80% of the odds, the result is within the range of €60 and €95 million.

The most probable result is equal to €77,160,743, an average of the 2109 results obtained with random combinations of values included in the interval of variability of the parameters characterised by greater uncertainty.

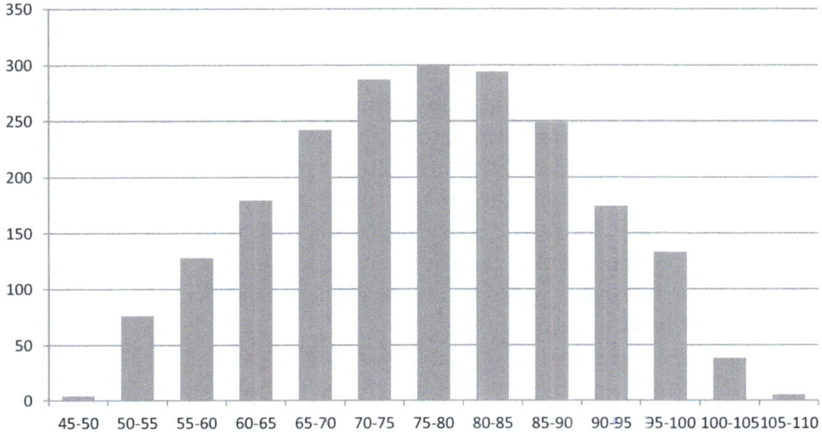

Fig. 4 Probability distribution of the NPV

5 Conclusions

In this paper, we offered a new perspective on the external costs generated by air transport services. In particular, we focused on how passengers arrive at and leave the airport, and we argued that some of them would switch from car, taxi or bus usage to the train if provided with implicit incentives. The type of incentive we considered is the introduction of a smart payment tool based on the credit card.

We first conceptualised the mechanisms behind the generation of changes in social welfare and then considered, from an empirical point of view, the case of the Malpensa Express connecting the centre of Milan to Malpensa Airport. The results of the cost-benefit analysis indicate that investing in the new technology generates positive welfare variations in a wide range of situations and that a significant portion of social benefits involve reductions in environmental externalities related to pollution and congestion.

Overall, the discussion of the introduction of a relatively minor technology has highlighted how consumers are very sensitive even to non-monetary incentives related to the ease of payments. In our case, a significant reduction in the external cost of mobility generated by trips between the airport and one's residence can be obtained by introducing a smart payment tool on the trains.

We live in a world where credit card-based apps make it easier to pay for services in competition with the railways, such as taxis or Uber cars. Given this fact and to avoid an increase in pollution around airports because of the diffusion of apps, we think that such payment systems need to be implemented in the railway lines connecting airports to cities.

Our findings are broadly in line with the sustainable aviation paradigm. Sustainable Aviation Network (2015) proposed a set of six goals to increase sustainability in the air transport sector: socio-economic, climate change, noise, local air quality, surface access, natural resources and implementation. The increase in the share of passengers reaching the airport by public transport will serve these objectives, especially in terms of improvement in local air quality.

In this view, aviation sustainability takes a multi-layer form as it involves not only carriers and their need for technological change (e.g. by using the less polluting planes and/or alternative fuels), but also airport operators (e.g. by improving energy efficiency of buildings) and finally users (passengers). This multi-layer strategy involves several actors and stakeholders that need to coordinate once sustainability strategies at airport level are drafted.

With specific reference to the case of passengers and their way to reach and depart from the airport, a crucial role is played by their preferences and by their behaviour (Ryley et al. 2013). Classical tools to deal with such issues are pricing roads (or the most polluting alternative) and expanding the supply of public transport services. In the case described in this paper, we have demonstrated that a substantial gain may come from increasing accessibility not through supply expansion, but by making the transport experience 'easier'. This policy opens the road to considering investment with modest cost and relatively significant social rate of return, which may be of particular interest for airports with medium capacity and lower financial space for investment.

Acknowledgements This chapter relies on a research project funded by Mastercard, whose financial support is gratefully acknowledged. The research was conducted by a team composed of Oliviero Baccelli, Giuseppe Siciliano Ugo Tortorì, Roberto Zucchetti and the author. The author affirms that Mastercard did not receive advance knowledge of the final outcomes of this research and at no point influenced the results of this study. The author would further like to add that even though this study focuses on smart payment systems linked to credit cards, there are various alternatives that utilise, e.g. hand-held devices for smart payment transactions and offer the same benefits described in this study.

References

Bell, N. (2003). *Development policy as public finance*. Oxford: Oxford University Press.

Comune di Milano. (2016). *Piano Urbano della Mobilità Sostenibile*. Milano: AMAT e Comune di Milano.

European Commission. (2014a). *Guide to cost-benefit analysis of investment projects*. Bruxelles: DG Regio.

European Commission. (2014b). *Update of the handbook of external costs of transport*. Bruxelles: DG Move.

Karakoc, T. H., Colpan, C. O., & Sohret, Y. (Eds.). (2018). *Advances in sustainable aviation*. Heidelberg: Springer.

Kaszewski, A. L., & Sheate, W. R. (2004). Enhancing the sustainability of airport developments. *Sustainable Development, 12,* 183–199.

Percoco, M. (2016). L'introduzione di strumenti di pagamento contactless: un'analisi costi-benefici per Milano. *Rivista di Economia e Politica dei Trasporti* (3), articolo 2.

Quibria, N. (2008). *The contactless wave: A case study in transit payments*. Federal Reserve Bank of Boston, Mimeo.

Ryley, T., Elmirghani, J., Budd, T., Miyoshi, C., Mason, K., Moxon, R., et al. (2013). Sustainable development and airport surface access: The role of technological innovation and behavioral change. *Sustainability*, 5(4), 1617–1631.

Sustainable Aviation Network. (2015). *Sustainable aviation. A decade of progress*. London: Sustainable Aviation Network.

CHAPTER 14

Sustainable Alternative Air Transport Technologies

Andreas B. Hardeman

1 Introduction

Throughout its history, aviation has been characterized by the gradual yet unrelenting innovation and adoption of new technologies and business paradigms. As in other areas of society, technology is driving change in the aviation sector at a ferocious pace, disruption has become the norm and the established order is constantly being challenged to adapt and transform.

New technologies are invented constantly to overcome the barriers of older technologies thus allowing for technological cross-fertilization and an acceleration of innovation to an ever-greater degree. Whether or not one subscribes to Kurzweil's *Law of Accelerating Returns*—according to which the rate of technological change tends to increase exponentially to the point where machine intelligence will surpass human

The views expressed in this paper are those of the author alone and do not necessarily represent those of any business affiliates or clients

A. B. Hardeman (✉)
Jurist in International Air and Space Law
and Independent Consultant, Geneva, Switzerland
e-mail: info@hardeman.consulting

T. Walker et al. (eds.), *Sustainable Aviation*,
https://doi.org/10.1007/978-3-030-28661-3_14

intelligence[1]—empirical observation of recent technological advancements in the aviation and aerospace sectors seems to justify the impression that things are indeed changing at an unprecedented rate.[2]

Within the last decade alone, the world has seen the successful introduction of sustainable biofuels for aviation, the proof of concept for solar-powered flight, the 3D printing of aircraft parts, the renaissance of battery-powered flight, the commercialization of suborbital spaceflight, and the emergence of autonomous air taxis and cargo drones, while supersonic airplanes are about to make a comeback, only this time without the much-maligned sonic boom. Beyond the realms of civil air passenger and cargo transport, and thus outside the scope of this chapter, equally impressive developments are underway involving hypersonic transport systems, and a wide array of (often solar-powered) high-altitude platforms and pseudo-satellites, including airships and balloons for remote sensing, surveillance and telecommunication purposes.

At the turn of the millennium, none of these technologies were considered to be practically feasible, or even possible.

The extent to which each of these new air transport technologies is expected to make a sustainable contribution to society will be investigated along with a discussion of regulatory requirements to test their environmental sustainability in terms of CO_2 emissions and noise performance. Following an overview of some key definitional, jurisdictional and standard applicability aspects in Sects. 2 and 3, the remainder of this chapter will take a closer look at cargo drones (Sect. 4), (hybrid-) electric flight (Sect. 5), and high-altitude, supersonic flight (Sect. 6).

2 Aircraft Classification

The applicability of the legal regime(s) that will ultimately govern the activities undertaken by these "newcomers" will be determined by their respective technological and operational specifications. A central question in this regard relates to the international standard-setting powers of the International Civil Aviation Organization (ICAO).[3]

[1] Kurzweil (1999).

[2] Some critics of the idea of exponential technological change have dismissed it as a collective "fetish of the second derivative" (Wichmann 2018).

[3] ICAO is one of the largest specialized agencies in the United Nations family, with 191 member states and an unparalleled law-making authority over 72% of the earth's surface.

Among ICAO's aims and objectives, as stated in Article 44 of its founding document the 1944 Chicago Convention (hereafter "the Convention"), are to foster the planning and development of international air transport so as to ensure the safe and orderly growth of international civil aviation throughout the world, encourage the arts of aircraft design and operation for peaceful purposes, and promote generally the development of all aspects of international civil aeronautics.

It may come as a surprise, therefore, that while the term "aircraft" appears more than 100 times in the Convention, it does not contain a definition of the term. This situation was only remedied in 1967 when Annex 7 to the Convention was amended to include the definition of an aircraft as "*Any machine that can derive support in the atmosphere from the reactions of the air other than the reactions of the air against the earth's surface.*" Annex 7 goes on to specify that "*an aircraft which is intended to be operated with no pilot on board shall be further classified as unmanned,*" and that "*unmanned aircraft shall include unmanned free balloons and remotely piloted aircraft.*"

As a consequence of this definition, it would appear that ICAO's remit, as circumscribed in the Convention, does in principle extend to the operation of cargo drones, electrically powered aerial vehicles as well as those operating at higher altitudes, such as super- and hypersonic machines, at least based on the designs currently under consideration. The designation as "aircraft" is not, as such, directly linked to the velocity, propulsion source or the altitude of the "machine"[4] (see Fig. 1).

More complicated would be the coverage of aerospace vehicles used for (sub)orbital flight that may use different techniques for different flight phases to "*derive support in the atmosphere.*" NASA's Space Shuttle, for instance, operated like a (vertical) rocket during launch and ascent

[4]With respect to altitude, this is the case as long as the "machine" in question "*can derive support in the atmosphere from the reactions of the air other than the reactions of the air against the earth's surface.*" Currently, an altitude of 100,000 ft. is considered the physical limit for "air breathing" airplanes. This ties in with a broader and more fundamental debate that has been raging in academic circles for decades regarding the question as to whether aerospace activities should be governed by principles of Air Law or Space Law. The two leading, opposing schools of thought respectively embrace the "spatialist" and the "functionalist" approach to determine applicable law. As neither of these two approaches provides us with an unambiguous framework, some have proposed alternative approaches (Dempsey 2018).

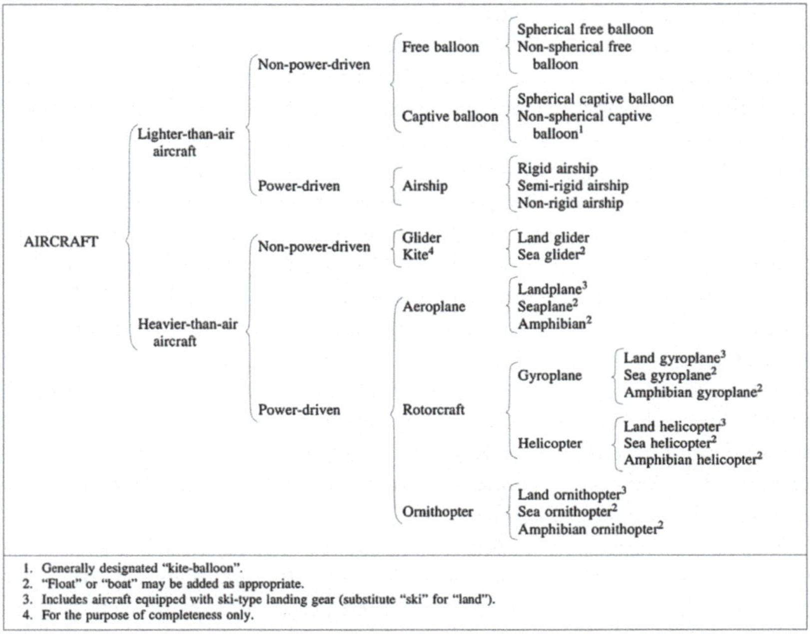

Fig. 1 ICAO 1944 Chicago Convention Annex 7—classification of aircraft

but as a glider during the descent and landing phases.[5] Similarly, ICAO jurisdiction over emerging air transport systems, such as those being operated by Virgin Galactic, Blue Origin, and SpaceX, will be a function of whether these systems fall within the ICAO definition of an "aircraft." According to Dempsey and Jakhu, ICAO does have the authority to amend Annex 7 to widen the definition of an "aircraft" to include aerospace vehicles of this type, but has not done so yet.[6]

[5] Note that these did not fall under ICAO jurisdiction in the first place due to not being *civilian* aircraft (cf. art. 3a of the Chicago Convention). International Civil Aviation Organization (ICAO 2006).

[6] P. S. Dempsey (2010) and Jakhu (2011).

3 ICAO Noise and CO_2 Standards

While acknowledging that environmental sustainability of flight operations encompasses a much wider range of impacts, the exclusive focus of this chapter is on aircraft-related noise and CO_2 emissions. It should be noted carefully at this point, however, that the discussion here is about the *applicability* of ICAO Standards and Recommended Practices (SARPs)[7] and not an attempt to prejudge or predict pass/fail probabilities *vis-à-vis* these standards of future aircraft designs.

3.1 ICAO Noise Standard

ICAO's aircraft noise certification standards are published in Annex 16, Volume I of the 1944 Chicago Convention. According to ICAO, the prime purpose of aircraft noise certification is "*to ensure that the latest available noise reduction technology is incorporated into aircraft design demonstrated by procedures which are relevant to day to day operations, to ensure that noise reduction offered by technology is reflected in reductions around airports.*"[8]

ICAO continually monitors research and development in noise reduction technology, complementing the standard-setting process.[9] Periodic reviews have led to increasingly more stringent standards over time in line with technological advancements. Each new standard is issued as a new chapter. As shown in Fig. 2, Chapter 14, the most recent noise standard adopted in 2013, followed three earlier standards: Chapters 2 (1973), 3 (1977), and 4 (2001).[10]

The noise metric used for certification purposes is Effective Perceived Noise in Decibels (EPNdB), which is a measure of human perception of aircraft noise that factors in sound intensity, duration of the noise event as well as the sound frequency spectrum and content of (particularly

[7] SARPs are technical specifications adopted by the Council of ICAO in accordance with Article 37 of the Convention on International Civil Aviation in order to achieve "*the highest practicable degree of uniformity in regulations, standards, procedures and organization in relation to aircraft, personnel, airways and auxiliary services in all matters in which such uniformity will facilitate and improve air navigation*" (ICAO 2006).

[8] ICAO (2006, 2007).

[9] See, for example, ICAO (2014a).

[10] The jump from Chapter 4 to Chapter 14 happened as a result of Chapter 5 being already used for a different standard and the next available Chapter was 14.

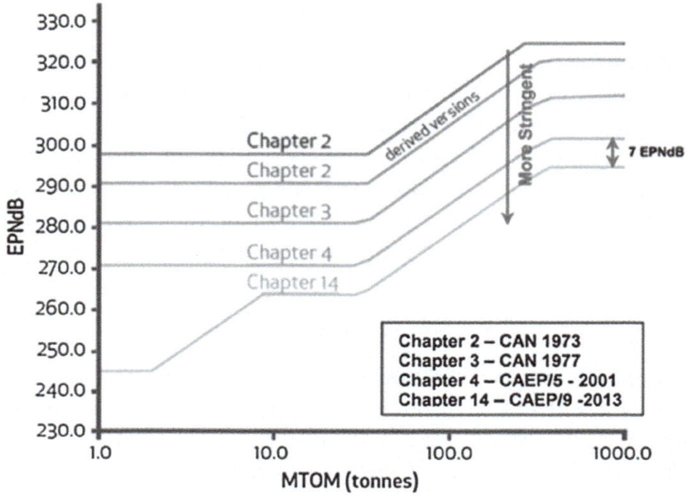

Fig. 2 Progression of ICAO noise standards (*Source* ICAO)

annoying) single tones. The EPNdB metric was specifically developed for aircraft noise certification.

Annex 16, Volume I noise standards apply specifically to "aeroplanes," defined in Part 1 as: "*A power-driven heavier-than-air aircraft, deriving its lift in flight chiefly from aerodynamic reactions on surfaces which remain fixed under given conditions of flight.*" An exclusion is granted to, *inter alia,* aeroplanes having short takeoff and landing (STOL) capabilities. Pending the development by ICAO of a suitable definition, these aeroplanes are described, for the purpose of this Annex, as those requiring a runway (with no stopway or clearway) of 610 m or less at the maximum certificated mass for airworthiness.

The Chapter 14 noise standard applies to new larger subsonic aeroplane type designs with a maximum certificated takeoff mass (MTOM) of 55 tonnes submitted for certification on or after December 31, 2017. For smaller subsonic aeroplanes with a MTOM of less than 55 tonnes, the standard will apply on or after December 31, 2020.

In February 2019, ICAO's Environmental Committee agreed on new noise technology goals, including improvements of aircraft noise up to 15.5 dB below Chapter 14 limits for single-aisle aircraft by 2027.

3.2 ICAO CO_2 Standard

In 2017, following a decade of intense deliberations, the ICAO Council adopted the world's first global design certification standard governing aircraft CO_2 emissions. In its essence, the standard is based on the maximum specific air range (SAR), which measures the distance (km) traveled per mass (kg) of fuel under optimal flight conditions. It can thus be taken as a measure of cruise fuel efficiency.

The metric value (MV) is expressed as $1/(SAR \times RGF^{0.24})$, where RGF is the reference geometric factor used to approximate useable aircraft space and includes all pressurized space on the main or upper deck(s) except the cockpit crew zone. The full standard is contained in a new Volume III to Annex 16 of the 1944 Chicago Convention.

The CO_2 standard described in Volume III will be applicable to subsonic aeroplanes with a MTOM of over 5.7 tonnes. In practice, this will cover all commercial jet aircraft (passenger and cargo) and all except the smallest of business aviation jets. Those weighing more than 60 tonnes, and thus having higher absolute emissions, will be subject to higher stringency levels, whereas smaller aircraft, with less potential for aerodynamic and weight efficiencies, will be subject to lower stringency levels. Volume III also applies to all propeller-driven aeroplanes with a maximum certificated takeoff weight of over 8.6 tonnes. Again, this includes all but the smallest commercial propeller aircraft.

Further, all new type (NT) aeroplanes (for which manufacturers will apply for a "type certification" after January 1, 2020) are covered[11] whereas all models that are currently in production (InP) will have to meet the slightly less stringent InP CO_2 standard after 2023 if they undergo modifications that require re-certification. The production of aeroplanes that do not meet the CO_2 standard will have to be discontinued from 2028. Figure 3 provides a graphic illustration of the applicable CO_2 standard stringency levels under Annex 16, Volume III.

ICAO will periodically review aircraft technology and assess the stringency level of the CO_2 standard, with the first technology review to be concluded in 2019. ICAO member states have explicitly recognized that the standard "is a technical comparison of aviation technologies designed for use in CO_2 emissions certification processes" and was "not designed

[11] Volume III grants special treatment to new type aeroplanes with a capacity below 19 seats, which will be covered from 2023.

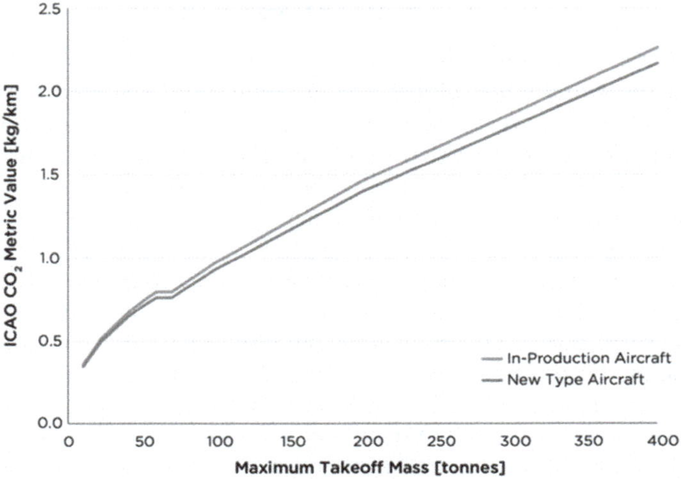

Fig. 3 Annex 16, Volume III CO_2 standard stringency levels (*Source* ICCT 2018)

to serve as a basis for operating restrictions or emissions levies",[12] something that may curb the incentivizing potential of the CO_2 standard.

Interesting to note is that amphibious aeroplanes are listed among the exceptions formulated in Annex 16, Volume III, along with, *inter alia*, certain aeroplanes conducting scientific, research, or humanitarian missions, as well as aeroplanes initially designed or modified and used for "specialized operational requirements," for example those equipped with a ramped cargo door.[13]

The following sections will present an overview of the selected new generation air transport technologies along with a brief assessment of their environmental characteristics and the applicability of existing ICAO noise and CO_2 standards.

[12] ICAO (2016).

[13] Examples of specialized operational requirements include: (a) aeroplanes that are initially certified as civil aeroplanes during the production process but immediately converted to military aeroplanes; (b) a required capacity to carry cargo that is not possible by using less-specialized aeroplanes (e.g., ramped, with back cargo door); (c) a required capacity for very short or vertical takeoffs and landings; (d) a required capacity to conduct scientific, research, or humanitarian missions exclusive of commercial service; or (e) similar factors.

4 Cargo Drones

Drones of all shapes and sizes have been widely touted as a "disruptor" and a "game changer."[14],[15] While industry initially underestimated their utility, drones are starting to be used in a wide spectrum of transport activities, from e-commerce package delivery, to transport of medicines, to fleet management and spare-parts delivery and even to same-day food delivery.

Several leading Big Tech and mail delivery giants are preparing to launch package delivery services using drones, including Amazon, Alphabet/Google, UPS, and Deutsche Post DHL. The use of drones for commercial package delivery is expected to become part of a new global market for drone-powered business solutions valued at around US$ 127bn.[16]

As can be seen from Fig. 4, drone systems (more formally called "unmanned aerial systems" [UAS]) can be grouped into five categories according to their field of application, namely surveillance, operations, entertainment/advertising, signal emission, and movement.

Cargo drones, sometimes referred to as Unmanned Cargo Aircraft (UCA), have long been used in military applications, but have so far been flying under the radar when it comes to commercial applications.[17] Analysts believe that this is now about to change: "the $15.5 trillion global freight market is poised for disruption" by seeking to fill a consumer niche between slow boats and pricy aircraft.[18] In more affluent parts of the world, interest in cargo drones has focused on the "last mile" but, as identified by experts at the World Economic Forum (WEF), the bigger opportunities exist in parts of Africa, Asia, and Latin America where they can augment existing distribution networks in remote regions "where poverty and disease are pervasive, distances are great, and roads will never be built."[19]

[14] Lawson (2017), Consumer Technology Association (CTA 2016), and Segarra (2018).

[15] As with many so-called new and advanced technologies, the "drone revolution" we are witnessing is really a continuation of a concept that has been around for a long time. The first successful example of a remotely piloted flight was the de Havilland DH82B *Queen Bee*, which entered service in Britain in 1935.

[16] PwC (2016).

[17] In the context of this chapter, cargo drones are distinguished from their smaller cousins, the individual "package delivery drones."

[18] The Yale Tribune (2017).

[19] Ledgard (2015).

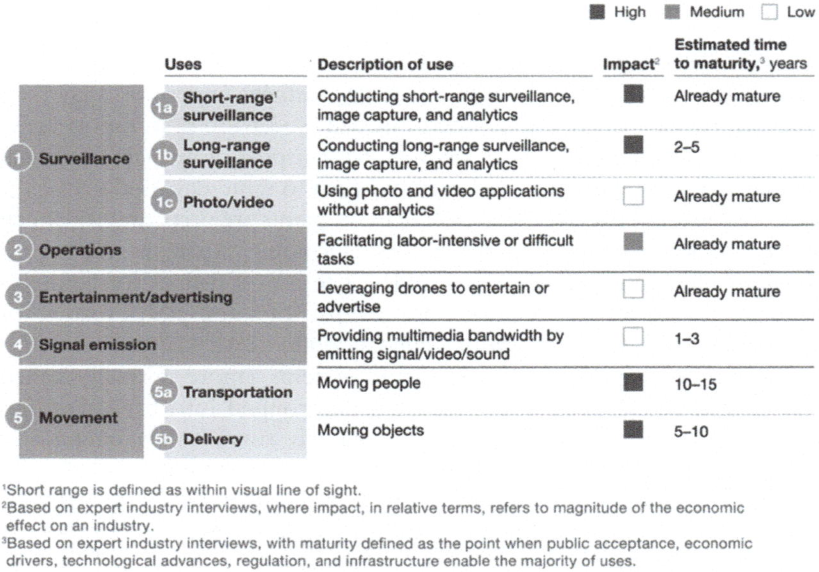

Fig. 4 Five categories of UAS applications (*Source* McKinsey 2017)

In September 2018, a company called Singular Aircraft completed its first major, long range, northern hemisphere autonomous flight test of the amphibious *FlyOx* drone at the Andoya NATO Air Base in Norway. The flight lasted two and a half hours and covered 440 km. The FlyOx is the world's largest amphibious UAV with a payload of 1850 kg and endurance of up to 25 hours. Kenyan operator Astral Aerial Solutions (AES) announced plans to launch its first cargo operation with the FlyOx drone in Africa by the end of 2018.[20]

In China, Tengoen has partnered with delivery company SF Express to build a high-capacity version of its *TB-001 Scorpion* cargo drone. It has a maximum takeoff weight of 3000 kilograms with a range of more than 6850 kilometers. The same company is also building an eight-engine drone with a wingspan of more than 41 meters to carry a payload of 18,000 kilograms up to 8630 kilometers, similar to a medium-sized manned cargo plane (e.g., the Boeing 737-700C).[21]

[20] Bekele (2018).

[21] Lin (2018).

In the USA, California-based start-up company Natilus has launched a prototype amphibious cargo drone capable of carrying 1800 kilograms over 1850 kilometers. If successful, the design will be scaled up to a 42-meter drone with a 90,000 kilogram cargo capacity that, once certified, could be flying routes between the USA and China in the foreseeable future.[22] This would place the drone firmly within markets currently served by wide-body freighter aircraft such as the McDonnell Douglas MD-11, Airbus A330-200F, and the Boeing 777 Freighter.

Considering its many advantages, it is not difficult to see why the prospect of UCA is quickly gaining popularity. Research by the Platform for Unmanned Cargo Aircraft (PUCA)[23] sums it up as follows:

- Duty length of a crew is not an issue, so UCA can have a cruising speed that is optimized to consume as little fuel as possible. The optimal speed will mostly be around 450 km/h, so a trip from eastern China to Paris will take one day. This is much faster than other forms of transport except manned aircraft, and fast enough for most cargo.
- The low cruising speed allows a light and efficient construction, the use of efficient propeller propulsion, and it makes the use of short, unpaved, runways possible. Also, the reduced fuel consumption yields greater range than comparable manned aircraft.
- Because there is no need for a pressurized crew cabin, a UCA can be made lighter and simpler than manned aircraft while still being able to fly efficiently at altitudes up to at least 6 km. Because the cross-section of the fuselage does not need to be circular, as is the case with a pressurized cabin, it can be shaped efficiently, for example to fit the shape of square cargo containers.
- The cargo area can be relatively small because no humans need to be accommodated. This gives the opportunity to use shapes like a Blended Wing Body (BWB) or flying wing, which is 15–20% more aerodynamically efficient than a conventional aircraft shape.

The US FAA predicts that in forty years' time, some 40% of air cargo will be transported by unmanned aircraft, or 25 million metric tonnes in today's terms.

[22] Freeman (2017).

[23] University of Twente (2018).

4.1 *Sustainability and Regulation of Cargo Drones*

What is commonly referred to as a drone is known as an "unmanned aircraft" (UA) in international legal terms. ICAO has defined a UA as "an aircraft which is intended to operate with no pilot on board."[24] This means it is either remotely and fully monitored and controlled from another place (ground, another aircraft, space), in which case it is typically referred to as a "remotely piloted aircraft" (RPA), or it is programmed and fully autonomous. All UA, whether piloted remotely, fully autonomous or a combination thereof, are considered "pilotless aircraft" in the sense of Article 8 of the Chicago Convention, which is aimed at avoiding dangerous interference with civil aircraft.[25]

Correspondingly, the current regulatory focus with regard to drones has been predominantly on safety and interaction with manned aviation. But like manned aircraft, cargo drone operations will have an impact on the environment depending on their size and weight, the type and amount of fuel consumed, and the nature and location of the operation, which must be taken into consideration. The International Transport Federation (ITF), for instance, identified the following areas as requiring regulatory and policy guidance: noise, vibrations, and light pollution and their impact both on humans and wildlife, energy consumption, emissions, and overall sustainability of drone transport. Among these, it has singled out the impact on noise levels, especially in urban and peri-urban settings, as one of the most important issues associated with a more widespread use of commercial drones.

To date, no comprehensive expert reviews have been conducted to determine the precise need for cargo drone noise and emissions control, although some sporadic comparative research exists. ICAO Circular 328 on Unmanned Aerial Systems (2011), for example, makes mention of a comparison between a small single-engine, manned aeroplane and a small UA showing that "the UA will, on average, use a tenth of the fuel, produce a tenth as much CO_2, have noise levels 6 to 9 dB lower, and operate five to ten times as long." Taking a more holistic, life-cycle approach, Stolaroff et al. (2018) found that small package deliveries with small drones have lower emissions in grams of CO_2 equivalent per package than trucks, but for larger packages, it is probably better to use electric trucks than large drones.

[24] ICAO (2011).

[25] ICAO (2011).

On the one hand, since electric drones can presently only cover short distances, larger cargo drones will be powered—at least for the foreseeable future—by conventional combustion engines which produce exhaust emissions and will be relatively noisy. On the other hand, they can be made lighter and smaller than the aircraft currently being used for many operations, which would generally make them more fuel efficient, less carbon intensive, and less noisy than their manned counterparts.

Current ICAO environmental standards regarding noise, engine emissions, and CO_2, as specified in Annex 16, Volumes I, II, and III respectively, are expected to apply to cargo drones assuming conventional airframe and engine designs are used. If revolutionary new designs were to be introduced, additional standards may have to be developed. Separate consideration may be warranted for larger electric cargo drones, for example, due to design characteristics that are unique to electric aircraft and their superior performance in terms of noise and CO_2 emissions.

An additional issue may be related to the fact that application of the CO_2 Standard relies on the "reference geometric factor" (RGF, see Sect. 3), which is well defined for a conventional "tube-and-wing" aircraft with a standard cabin, but would probably yield meaningless results for more exotic configurations, for example BWB designs, as may be the case for the larger cargo drones currently under development. This means a modified CO_2 metric value formula may have to be defined for this new category of aircraft in the future.

Finally, consideration must be given to the fact that Annex 16, Volume III excludes amphibious aircraft as well as aircraft with ramped, back cargo doors from the existing CO_2 standard. At present, this would give a free pass to some of the largest cargo drones under development.

International airlines, meanwhile, appear keen to facilitate the adoption of drones and make them an efficient and a viable alternative to conventional aircraft, emphasizing that manufacturers still need to improve safety-related technologies, such as geo-fencing and collision avoidance, and reduce noise.[26] The conversion of conventional aircraft into UCA is also being explored.

Airports, who are traditionally in the front line of responding to public aircraft-related environmental health concerns, have been calling for drones to "be subject to the same noise and emissions standards applied to

[26] International Air Transport Association (IATA 2018).

manned aircraft at an airport" and "be operated in compliance with published airport noise abatement procedures."[27] It should be recognized, however, that UA may or may not be operated out of traditional airports where compliance with noise standards would be required. Indeed, UA operations may occur in ad hoc or semi-prepared locations away from populated areas (sometimes referred to as "droneports"), raising questions as to whether operational noise requirements need necessarily be applied. A question raised in this particular context would be the effects that drone-related noise may have on land use values close to droneports.[28] Rather than waiting for local communities to introduce their own individual noise and fuel efficiency rules, it would be best for industry to proactively draw up voluntary protocols or even propose international rules in this area.

5 Solar and (Hybrid-) Electric Flight

Despite the undisputed potential of solar energy for many applications, solar-powered flight is not one of them, at least not yet. As much as Solar Impulse 2 was decidedly an amazing feat of dare and engineering, team leaders Piccard and Borschberg were quick to emphasize that solar-powered options will not be heading to commercial airlines anytime soon.[29]

Battery-powered flight, however, has a rich history. It was Gaston and Albert Tissandier who flew an electrically powered dirigible in 1883, two decades before the Wright Brothers' glider took to the air.[30] In recent years, the quest for zero-emission flight has taken off in all seriousness, with both established aerospace behemoths and new age garage start-ups launching ambitious projects.

Boeing and JetBlue have backed Zunum Aero to develop a family of 10- to 50-seat hybrid-electric regional aircraft while Airbus, Rolls-Royce, and Siemens are working on the hybrid-electric E-Fan X demonstrator, based on a BAE 146 RJ100 platform with one of the aircraft's four engines replaced by a 2 MW electric motor.[31] In a truly global effort,

[27] Airports Council International (ACI 2018).

[28] International Transport Forum (ITF 2018).

[29] Wei-Haas (2018).

[30] Swopes (2017).

[31] Airbus (2017).

ATR and Air New Zealand are exploring opportunities and implications of hybrid and electric aircraft on the regional aviation ecosystem.[32]

In Israel, Eviation is currently underway to manufacture the first prototype of its all-electric business and commuter aircraft, Alice. The all-composite aircraft, powered by a 900 kWh lithium-ion battery pack and 260 kW Siemens electric motors, will seat up to nine passengers and two crew in its 12 m long fuselage. It is projected to have a range of 540 nautical miles (nm) (1000 km) and a cruise speed of 240 kt (440 km/h) with a maximum payload of 1250 kg and a maximum take-off weight of 5900 kg.

In the USA, recent start-up Ampaire is developing a retrofitted electric aircraft with the aim to be FAA certified by the end of 2020. The aircraft will be able to carry 7–9 passengers and have a range of up to 100 miles.[33] Also in the USA, a company called Wright Electric aims to create a commercial airliner that runs on batteries for distances of less than 300 miles. Airline partner EasyJet announced all of its flights under 540 km to be zero emissions in 10–20 years.[34]

Meanwhile, German research institute Bauhaus Luftfahrt has been studying the Ce-Liner, an electric concept aircraft with a capacity of 180–200 seats and a design range of 1000–2500 km, depending on the desired entry into service of 2030, 2035, or 2040.[35] Special mention should also be made of Joby Aviation and Lilium, both looking to revolutionize individual air travel by developing unique all-electric vertical-takeoff-and-landing (eVTOL) vehicles with surprising capabilities (speeds of up to 300 km/h over a range of 300 km)—all on a single charge.[36]

The predominant constraining factor in entrepreneurs' ambitions remains the storage capacity (energy density) of available batteries. The best performing lithium-ion batteries currently available sport gravimetric energy densities of around 270 Wh/kg. By comparison, the energy density of jet fuel is around 12,000 Wh/kg.[37] The energy efficiency of

[32] Bradley (2018).

[33] Smith (2017).

[34] Smout (2017).

[35] Bauhaus Luftfahrt (2017).

[36] Lilium (2018).

[37] Janek (2016).

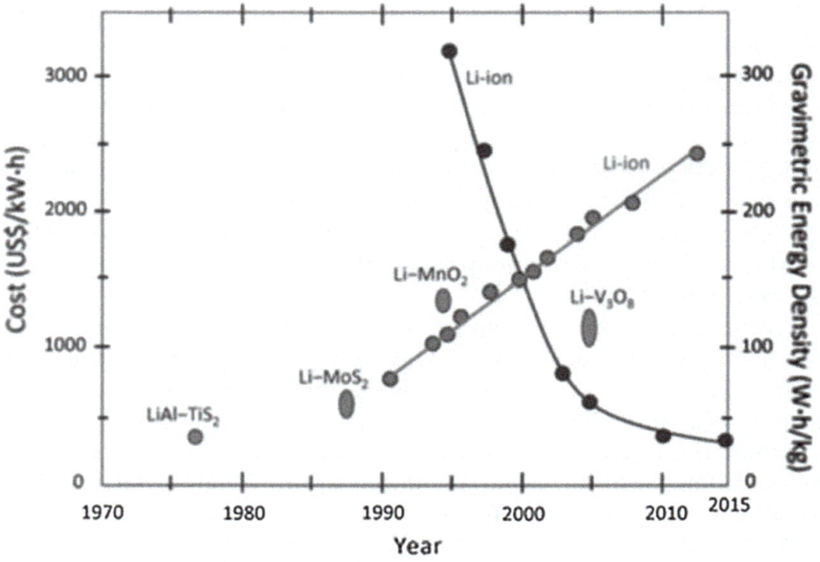

Fig. 5 Development of lithium batteries 1970–2015 (*Source* Crabtree et al. 2015)

electric engines is significantly higher than that of combustion engines, largely because electric engines generate motion instead of heat. Future generations of batteries, such as the lithium-oxygen battery, could reach an energy density of 400 Wh/kg, which according to experts would enable flights of 300–600 km (Fig. 5).[38]

Some have expressed the ambition to create zero-emissions aircraft designs, hoping to "reduce, and possibly eliminate, the use of fossil fuels in aviation within the next 30 to 40 years."[39] Norway aims to have all short-haul flights—those that last less than 1.5 hours—using electric planes by 2040.[40] In turn, this has already led to visions of dedicated "e-ports" built close to city centers to accommodate small hubs for electric "air buses" with 10–12 seats, up to maybe 40 seats,[41] taking

[38] Adams (2017).

[39] Weber (2017).

[40] Agence France-Presse (2018).

[41] Deutsche Welle (2018).

advantage of the fact that electric aircraft are expected to be substantially quieter than conventional aircraft.

5.1 Sustainability and Regulation of (Hybrid-) Electric Aircraft

The successful commercialization of electric flight will depend on many factors, not least the investment requirements and resource availability for battery production, storage, and end-of-life disposal. When considering the sustainability and regulatory requirements of electric flight, hybrid-electric aircraft (HEA) must be distinguished from full electric aircraft (FEA). Hybrid-electric aircraft have gas turbine engines which drive electrical generators to power electrically motor driven fans. Thrust may be provided via a combination of gas turbines and electrical propulsors, or only via the electrical propulsors. Noise and emissions will generally be lower as reliance on electrical propulsion increases. Further, as with all electrical systems, overall sustainability will in part rest on the associated life-cycle emissions, depending on how the electricity was sourced.

When it comes to emissions and noise performance, the consensus appears to be that technological advances in hybrid-electric flight hold great promise for minimizing aviation's impact on our environment as it helps to substantially improve overall efficiency, reduce air pollution, and decrease noise levels. Some believe the potential exists for up to 40% of all commercial aviation emissions to be eliminated by the 2030s, based on the assumption that the majority of short-haul, regional aircraft (about 8000 according to a 2013 Embry-Riddle study[42]) are due to be replaced by 2030.[43]

The Airbus A320neo, slated to become the first airplane to use the Wright Electric engine, is projected to produce 15% lower CO_2 emissions and a 50% lower noise footprint on takeoff and landing. With a maximum range of 335 nm, it would be suitable for about 20% of EasyJet's flights.[44] Elsewhere, studies have indicated potential fuel savings from hybrid-electric propulsion of up to 50% for a microlight

[42] Guzhva (2013).

[43] Brown (2018).

[44] Forsdick (2018).

aircraft and 10% for an inter-city airliner, respectively.[45] Benefits have equally been identified for ground operations, as evidenced by the development in recent years of electric taxiing systems powering the landing gear wheels, as well as towing tugs remotely controlled by the aircraft pilot.[46]

Reconfirming the projected noise benefits mentioned above associated with the Wright Electric engine, comparative research referenced by Sou (2018) shows a 14 decibel (dB) decrease in noise production.[47] In another example, sideline noise[48] for the Zunum electric aircraft line is around 65 EPNdB, which compares to 86.5 EPNdB for a Cessna 500 with an identical maximum takeoff weight (MTOW) of 5216 kg.[49] This represents a noise reduction of more than 20 EPNdB. Since decibels are measured on a logarithmic scale, a 20 EPNdB decrease in noise levels corresponds to about a 75% reduction in perceived noise.

From a regulatory perspective, it is worth noting that certification of electrically powered commercial aircraft is likely to have its own unique requirements, some of which will differ substantially from those included in existing certification procedures. Current FAA and EASA certification regulations for commercial aircraft presently do not address (hybrid-) electric design aspects associated with high-power propulsion systems, such as thermal effects, electric thrust management, altitude and dynamic effects on electric systems, and electromagnetic compatibility issues. In anticipation of these future needs, the E-Fan X demonstrator program is aiming to establish future certification requirements for electrically powered aircraft along with specialized training of a new generation of HEA designers and engineers.[50]

In this context, useful lessons can be learned from the successful international cooperation between EASA and the FAA in rewriting certification rules CS-23/Part 23,[51] providing for the airworthiness certification

[45] Friedrich (2014).

[46] Cummings (2009).

[47] Sou (2018).

[48] Defined as "the maximum noise level on a line parallel to the runway at a distance of 450 m to the runway centerline".

[49] European Aviation Safety Agency (EASA 2018).

[50] Airbus (2017).

[51] Applicable to aeroplanes with a passenger seating configuration of 19 or less and a maximum certified takeoff mass of 8618 kg (19,000 lb) or less.

of fully electric training aircraft.[52] The new CS-23 rules[53] establish objective and design-independent requirements, meaning new designs will not be hampered by detailed prescriptive rules. This enables innovative solutions to enhance safety while at the same time reducing administrative burden and certification costs. Although not a first (certification of small electric aircraft was already possible in Australia and Canada) the EASA/FAA collaboration is testament to what can be accomplished when governments and industry work hand in hand to achieve a common goal.

Overall, (hybrid-) electric aircraft would be expected to meet existing ICAO noise and CO_2 standards with comfortable margins, if these standards were to be applied. From a standard-setting perspective, however, it may be worth considering a separate set of ICAO standards for (hybrid-) electric aircraft to address the unique features associated with high-power propulsion systems. Separate standards would also avoid conventionally powered aircraft coming under pressure to meet technologically unfeasible noise and CO_2 limits.

At the time of going to print, ICAO had just announced plans to begin the assessment of how to certify new technologies, such as hybrid and electric aircraft, as parts of its future work program up to 2022.

6 High-Altitude (Supersonic) Flight

Reviving supersonic passenger flights is something many people have dreamed of ever since the Concorde was retired in 2003, and it is possible that we will see the first, new generation supersonic transport (SST) jets come into service within the next decade. The three US aerospace firms leading the SST renaissance—Aerion Supersonic, Spike Aerospace, and Boom Supersonic—while working on distinctive designs, all plan to have their aircraft in regular service by 2025.[54] A 2016 study by the Boyd Group, commissioned by Boom Supersonic, estimates that the market for supersonic aircraft could be 1300 units, worth US$ 260

[52]The first electric aircraft to receive its airworthiness certification from the FAA was Pipistrel's Alpha Electro, a 2-seat electric trainer aircraft. It has an all-composite body, electric motor and 20 kWh battery packs and weighs 350 kg with a maximum payload of 200 kg.

[53]European Aviation Safety Agency (EASA 2017).

[54]Bowler (2018).

billion, over a 10-year period.[55] Eventually, over 500 future global routes may benefit from supersonic travel, with transatlantic and transpacific routes being key.[56]

The technological challenge facing manufacturers of new supersonic equipment is to come up with an airframe-engine design combination that is able to simultaneously optimize cruise efficiency, mitigate the sonic boom effect during cruise as well as reduce noise and emissions during the landing and takeoff (LTO) cycle.[57]

Aerion Supersonic, GE Aviation, and Boeing are working together to produce the AS2 business jet, a 12-seat trijet, with a supersonic range of 4200 nm and 5400 nm at subsonic speeds.[58,59] At the time of writing, it had received 20 orders from fractional ownership operator Flexjet.[60] Considered to be the most advanced along the development curve, the AS2 does not incorporate any specific low-boom design features. Instead, it will rely on the "Mach cut-off" phenomenon whereby boom shock waves are refracted upwards as they enter warmer air closer to the ground. This would allow the AS2 to fly up to Mach 1.2 without a sonic boom reaching the ground.[61] The use of carbon fiber and new wing configurations may allow the jet to achieve supersonic natural laminar flow (SNLF), a proprietary technology developed by Aerion that the company says can reduce airframe drag by as much as 20% compared to the Concorde, thus creating significant potential for achieving fuel and CO_2 efficiencies.[62]

At the same time, Spike Aerospace is developing the S-512 "Quiet Supersonic Jet," a twin-engine business jet design aiming to reach Mach 1.6 cruise speed over water for 6200 nm with 18 passengers. The S-512 is scheduled to fly in 2021 and should have a boom no louder than

[55] Gray (2018).

[56] Leadbeater (2016).

[57] The LTO cycle used by ICAO covers four modes of engine operation, namely idle, approach, climb out, and takeoff, each of which is associated with a specific engine thrust setting and a time in mode.

[58] AIN Online (2018).

[59] Simple Flying (2019).

[60] Bean (2018).

[61] Warwick (2012).

[62] Matousek (2018).

75 dB on the ground, largely due to its small size, the use of lightweight materials and its aerodynamic properties.[63]

Aiming for larger commercial markets, meanwhile, Boom Supersonic is anticipating the launch of a 55–75-seat trijet passenger aircraft, the *Overture*, by 2023, reaching Mach 2.2 over water for 9000 nm.[64] Tests of the XB-1 Supersonic Demonstrator "Baby Boom" are scheduled for 2019.[65] At the end of 2018, five airlines had placed a total of 76 orders, including Virgin Group and Japan Airlines.[66]

In parallel with the planned launch of these new supersonic jets, cutting-edge technologies are being developed by NASA and Lockheed Martin as part of their low-boom flight demonstrator (LBFD, aka X-59 Quiet Supersonic Transport [QueSST]). Designed to cruise from 44,000 to 59,000 feet at a speed of at least Mach 1.4, it is aiming for envelope expansion flights in 2021 and demonstration flights to take place in 2022.[67]

Compared to a Concorde's N-wave sonic boom loudness of around 107 PLdB (89 dBA SEL),[68,69] the demonstrator avoids the convergence of individual pressure waves at supersonic speeds to achieve a far less audible shaped sonic boom of 75 PLdB (59 dBA SEL) in level flight.[70,71] The LBFD will assess varying loudness by producing shaped boom variations from 70 to 80 PLdB through altitude variations (amplitude), trimmed control deflection variations (shape), or flight in a shallow turn of less than 10 degrees in bank (varied carpet loudness).[72]

[63] Spike Aerospace (n.d.).

[64] Boom Supersonic (n.d.).

[65] Adams, The 'baby boom' charts a return to supersonic flight (2018).

[66] Burgess (2017).

[67] Yenko (2018).

[68] Perceived Level of Sound (Noisiness or Loudness), or PLdB, is proportional to the peak energy level reaching the ear during a flyover event and is typically used to measure the human reaction to sonic boom levels.

[69] Sound Exposure Level (SEL) is the constant sound level that has the same amount of energy in one second as the original noise event. It provides a measure of the physical energy of the noise event which takes into account both intensity and duration.

[70] German (2018).

[71] Lockheed Martin (2017).

[72] A shallow turn increases loudness on the inside of the turn and decreases loudness on the outside.

To avoid sonic boom while also minimizing takeoff noise, supersonic aircraft designs may ultimately have to incorporate variable-cycle, three-stream turbofan engines. The advantage of this engine design is that a high bypass ratio (BPR)[73] can be used during takeoff to reduce noise while the pilot could lower the ratio in flight to increase specific thrust and optimize supersonic cruise efficiency. Early supersonic aircraft designs, however, could use medium-bypass engines to meet operational requirements, which can be adapted from existing engines designs currently in use to power subsonic wide-body aircraft.[74]

6.1 *Sustainability and Regulation of High-Altitude (Supersonic) Flight*

History has shown that the viability of supersonic flight depends significantly on social acceptability and, hence, the impact on its environmental surroundings in the broadest sense. Beyond atmospheric effects, including global warming and ozone depletion, and human annoyance levels related to aircraft noise, consideration may also need to be given to specific human health impacts as well as effects on (marine) wildlife.

Like its predecessors, the new generation of commercial and corporate supersonic jets is designed to fly in the lower reaches of the stratosphere, around 44,000–60,000 ft. This compares to a typical cruise altitude of 28,000–43,000 ft. for subsonic aircraft.[75] Although the adverse effect from aircraft nitrogen oxide (NOx) emissions on atmospheric ozone concentrations tends to increase at higher altitudes, NASA research conducted since the IPCC's Report on Aviation and the Global Atmosphere[76] suggests that emissions from supersonic aircraft in the lower stratosphere pose a negligible risk to the ozone layer.[77]

In regard to the fuel economy, Concorde managed about 17 passenger-miles to the Imperial gallon,[78] which is 16.7 L/100 km per

[73]The bypass ratio (BPR) of a turbofan engine is the ratio between the mass flow rate of the bypass stream to the mass flow rate entering the core. High BPR engines tend to be more fuel efficient and less noisy.

[74]Boom Supersonic (2017).

[75]Although some business jets (e.g., the Gulfstream G450+) fly to 50,000 ft.

[76]Intergovernmental Panel on Climate Change (IPCC 1999).

[77]Wey et al. (2004).

[78]Davies (2015).

passenger (pkm). Future generations of supersonic jets would have to be–5 times more fuel efficient to equal today's industry average of around 3.5 L/100 pkm. Due to the technological and operational tradeoffs that exist between optimizing fuel economy on the one hand, and noise performance on the other hand, it is estimated that supersonic aircraft could be up to 20% less fuel efficient if they were required to meet the latest subsonic noise rules for landing and takeoff.[79] This conundrum has led some to propose that the noise standard for supersonic aircraft should be more lenient than for subsonic aircraft.[80]

Surprisingly perhaps, the development of new supersonic aircraft is happening in a regulatory near-vacuum. Although technically, supersonic aircraft engines are subject to the emissions standards[81] laid out in Part III, Chapter 3, of Annex 16, Volume II, ICAO's Committee on Aviation Environmental Protection (CAEP) in 2007 agreed that the current supersonic standard seems outdated and should not be applied to new engine projects.[82]

Further, there are currently no internationally agreed standards for supersonic aircraft fuel or CO_2 performance, LTO noise, or the acceptability of sonic boom. Many aviation authorities, including in the USA and Europe, simply prohibit supersonic flight over land, irrespective of any associated noise or emissions levels.[83] This poses obvious challenges for the successful development and commercialization of any new products.

While an argument could be made for applying existing subsonic noise and emissions standards in the interim to provide regulatory certainty and ensure environmental protection,[84] a more fundamental question remains whether it would be appropriate to compare supersonic to subsonic aircraft in the first place. Supersonic jets are structurally, aerodynamically, and operationally a completely different species of aircraft whose performance should arguably not be shoehorned into subsonic certification standards.

To provide the necessary regulatory guidance, much depends on national aviation authorities cooperating to develop workable proposals

[79] Garcia (2018).

[80] Dourado (2017).

[81] For hydrocarbons (HC), Carbon monoxide (CO), Oxides of nitrogen (NOx), and black smoke.

[82] ICAO (2014b, pp. 2–17).

[83] Red (2008).

[84] ICSA views on future supersonic transport engines and aircraft (ICSA 2018)

within ICAO. If timely proposals fail to materialize, however, this would pose a clear incentive for some national authorities to go it alone and propose local rules to pave the way for noise and emissions certification of supersonic aircraft domestically. In fact, the FAA already announced in 2018 that it intends to issue Notices of Proposed Rulemaking in 2019 to prepare for noise certification of supersonic aircraft and facilitate supersonic overland flight-testing in the USA.[85] As of early 2019, this scenario came a step closer with environmental experts at ICAO agreeing to undertake exploratory studies into SST operations while stopping short of deciding to develop specific LTO Standards for supersonic aircraft.

7 Conclusion

A new generation of air transportation modes is being born before our eyes. Whether their arrival will be welcomed by society at large will depend to a significant degree on their perceived contribution to—or at least in compatibility with—sustainable social and economic development. The available evidence presented in this chapter suggests there are plenty of reasons to be optimistic.

Emergence of small and large cargo drones holds the potential to open up new markets, connect remote regions in less developed parts of the world, and lower the cost of cargo transportation while simultaneously producing fewer emissions and less noise than conventional alternatives. Advanced battery technologies can pave the way for electric air shuttles servicing urban e-ports and electric air taxis efficiently connecting air passengers with city centers and conference venues. Innovative airframe and engine designs using carbon fiber and specialized alloys will steadily build toward a viable fleet of supersonic airliners operating across the globe within the accepted boundaries of sustainability and respecting international environmental standards.

Generally speaking, there is a discernable trend toward the increased specialization of transport modes, the mission-specific optimization of designs, and the operational capabilities. This can be expected to produce significant time and resource efficiencies, *inter alia,* by reducing infrastructure needs and unburdening existing infrastructure, such as highways and airports.

[85] Federal Aviation Administration (2019).

It may well be that many of the new technologies will start to converge and enable each other, pushing their development along faster than expected. If regulations are unable to keep pace with these developments, however, this could jeopardize the introduction of new sustainable transport modes, along with their associated socioeconomic benefits.

Yet, if these regulatory pitfalls can be successfully avoided, new air transportation systems have a massive potential to bring together and mutually reinforce all three aspects of sustainable development (economy, society, environment) in a way that can even be described as "super sustainability": not only would we be able to meet the needs of the present "without compromising the ability of future generations to meet their own needs",[86] but we would actually be able to *improve* their ability to do so. Based on current trends, future generations of aspiring air travelers can expect to enjoy air transportation that is simultaneously more integrated, more inclusive, and more sustainable than ever before.

References

Adams, E. (2017, May 31). The age of electric aviation is just 30 years away. *Wired*. Retrieved from https://www.wired.com/2017/05/electric-airplanes-2/.

Adams, E. (2018, November 15). The 'baby boom' charts a return to supersonic flight. *Wired*. Retrieved from https://www.wired.com/story/boom-aerospace-baby-boom-demonstrator-supersonic/.

Agence France-Presse. (2018, January 18). Norway aims for all short-haul flights to be 100% electric by 2040. *The Guardian—World*. Retrieved from https://www.theguardian.com/world/2018/jan/18/norway-aims-for-all-short-haul-flights-to-be-100-electric-by-2040.

AIN Online. (2018, July 25). Aerion expects to fly the first supersonic business jet by 2023. *AIN Online*. Retrieved from https://www.ainonline.com/aviation-news/video/aerion-expects-fly-first-supersonic-business-jet-2023.

Airbus. (2017, November 28). Airbus, Rolls-Royce, and Siemens team up for electric future Partnership launches E-Fan X hybrid-electric flight demonstrator. *Newsroom—Airbus*. Retrieved from https://www.airbus.com/newsroom/press-releases/en/2017/11/airbus–rolls-royce–and-siemens-team-up-for-electric-future-par.html.

Airports Council International (ACI). (2018, July). *Drones policy paper version 1.0*. ACI Policy Papers. Retrieved from https://aci.aero/wp-content/uploads/2018/08/ACIPolicyPaper_Drones_2018-1.pdf.

[86] The World Commission on Environment and Development (WCED 1987).

Bauhaus Luftfahrt. (2017). Der Ce-Liner: potenzielle Emissionsfreiheit im kommerziellen Luftverkehr. *Bauhaus Luftfahrt—Topthema*. Retrieved from https://www.bauhaus-luftfahrt.net/de/topthema/ce-liner/.

Bean, L. (2018, July 31). Aerion is bringing supersonic flight back ... And this time it's private. *Robb Report*. Retrieved from https://robbreport.com/motors/aviation/aerion-as2-could-become-worlds-first-supersonic-private-jet-eg18-2808106/.

Bekele, K. (2018, May 15). Astral aviation to launch Africa's first cargo drone operation. *AIN Online*. Retrieved from https://www.ainonline.com/aviation-news/air-transport/2018-05-15/astral-aviation-launch-africas-first-cargo-drone-operation.

Boom Supersonic. (2017, April 11). Why we don't need an afterburner. *Boom Supersonic*. Retrieved from https://blog.boomsupersonic.com/why-we-dont-need-an-afterburner-a4e05943b101.

Boom Supersonic. (n.d.). *Boom overture*. Boom Supersonic. Retrieved on 24 June 2019, from https://boomsupersonic.com/overture.

Bowler, T. (2018, July 20). Is killing the boom the key to supersonic air travel? *BBC News*. Retrieved from https://www.bbc.com/news/business-44795639.

Bradley, G. (2018, November 9). Air New Zealand working with ATR to develop hybrid electric planes. *NZ Herald*. Retrieved from https://www.nzherald.co.nz/business/news/article.cfm?c_id=3&objectid=12157487.

Brown, A. (2018, November 21). Electric aviation may slash emissions by early 2020s. *Triple Pundit*. Retrieved from https://www.triplepundit.com/2018/11/electric-aviation-emissions/.

Burgess, M. (2017, June 21). Supersonic plane startup Boom has 76 orders and expects flight in 2023. *Wired*. Retrieved from https://www.wired.co.uk/article/virgin-supersonic-travel.

Cohn, P. (2017, December). Commercial drones are here: The future of unmanned aerial systems. *McKinsey & Company*. Retrieved from https://www.mckinsey.com/industries/capital-projects-and-infrastructure/our-insights/commercial-drones-are-here-the-future-of-unmanned-aerial-systems.

Consumer Technology Association (CTA). (2016, March 31). Why drones are the ultimate disruptive technology. *Futuretech*. Retrieved from https://medium.com/engine-futuretech/why-drones-are-the-ultimate-disruptive-technology-989f10092a0.

Crabtree, K., Kócs, E., & Trahey, L. (2015, December). The energy-storage frontier: Lithium-ion batteries and beyond. *NY-BEST*. Retrieved from https://www.ny-best.org/sites/default/files/resources/CrabtreeKocsTrahey_TheEnergyStorageFrontierLithiumIonBatteriesAndBeyond_MRSBulletin40106715.pdf.

Cummings, K. (2009, December 6). Reducing runway emissions. *The Engineer*. Retrieved from https://www.theengineer.co.uk/issues/7-december-2009/reducing-runway-emissions/.

Davies, H. (2015, September 29). Is the supersonic Concorde set to make a comeback? *MRO Network*. Retrieved from https://www.mro-network.com/airframes/supersonic-concorde-set-make-comeback.

Dempsey, P. S. (2010). Space traffic management: A vacuum in need of law. In P. Kumar (Ed.), *Outer Space: Warfare and Weapons*.

Dempsey, P. S. (2018, March 29). Suborbital flights and the delimitation of air space vis-à-vis outer space: functionalism, spatialism and state sovereignty. *A/AC.105/C.2/2018/CRP.9*. Vienna: UN Committee on the Peaceful Uses of Outer Space - Legal Subcommittee.

Deutsche Welle. (2018, April 23). How soon till we all fly in electric planes? *DW Top Stories*. Retrieved from https://www.dw.com/en/how-soon-till-we-all-fly-in-electric-planes/a-43473916.

Dourado, E. (2017, October 27). Make America boom again. *Mercatus Center, George Mason University*. Retrieved from https://www.mercatus.org/publications/make-America-boom-again.

European Aviation Safety Agency (EASA). (2017, March 29). Certification Specifications for normal-category aeroplanes (CS-23). *EASA*. Retrieved from https://www.easa.europa.eu/sites/default/files/dfu/CS-23%20Amendment%205.pdf.

European Aviation Safety Agency (EASA). (2018, June 24). EASA certification noise levels. *EASA Environment*. Retrieved from https://www.easa.europa.eu/easa-and-you/environment/easa-certification-noise-levels.

Federal Aviation Administration. (2019, February 27). Fact sheet—Supersonic flight. *US DOT Federal Aviation Administration*. Retrieved from https://www.faa.gov/news/fact_sheets/news_story.cfm?newsId=22754.

Forsdick, S. (2018, November 6). These are the companies making electric planes—And making aviation greener. *Compelo*. Retrieved from https://www.compelo.com/companies-making-electric-planes/.

Freeman, D. (2017, April 3). These giant drones could seriously disrupt the shipping industry. *MACH*. Retrieved from https://www.nbcnews.com/mach/innovation/these-giant-drones-could-seriously-disrupt-shipping-industry-n741981.

Friedrich, C. R. (2014, December 30). Hybrid-electric propulsion for automotive and aviation applications. *Springer Link, CEAS Aeronautical Journal*. Retrieved from https://link.springer.com/article/10.1007/s13272-014-0144-x.

Garcia, M. (2018, May 30). Supersonic air travel may be on verge of return—Just don't expect it from N.Y. to L.A. *Forbes*. Retrieved from https://www.forbes.com/sites/marisagarcia/2018/05/30/why-faa-noise-rule-review-doesnt-mean-you-will-fly-supersonic-from-ny-to-la/#b95172b5e8c1.

German, K. (2018, August 25). Supersonic without the boom. *CNET*. Retrieved from https://www.cnet.com/news/supersonic-without-the-boom/.

Gray, A. (2018, February 15). Supersonic jets are making a comeback. *World Economic Forum (WEF)*. Retrieved from https://www.weforum.org/agenda/2018/02/supersonic-jets-are-making-a-comeback/.

Guzhva, V. C. (2013). *Market analysis for small and mid-size commercial turboprop aircraft*. Daytona Beach: Embry-Riddle Aeronautical University.

Intergovernmental Panel on Climate Change (IPCC). (1999). *Aviation and the global atmosphere*. WMO and UNEP. Geneva: Cambridge.

International Air Transport Association (IATA). (2018). Cargo drones—Drones for tomorrow's air cargo. *IATA—What We Do—Cargo*. Retrieved from https://www.iata.org/whatwedo/cargo/Pages/cargo-drones.aspx.

International Civil Aviation Organization (ICAO). (2006). Convention on International Civil Aviation—Doc 7300/9. *ICAO Uniting Aviation*. Retrieved from https://www.icao.int/publications/pages/doc7300.aspx.

International Civil Aviation Organization (ICAO). (2007). *CAEP—Committee on Aviation Environmental Protection Report of the Seventh Meeting (Doc 9886)*. Montreal: ICAO.

International Civil Aviation Organization (ICAO). (2011). ICAO Cir 328, Unmanned Aircraft Systems (UAS). *ICAO Documents*. Retrieved from https://www.icao.int/Meetings/UAS/Documents/Circular%20328_en.pdf.

International Civil Aviation Organization (ICAO). (2014a). Doc 10017, report by the second CAEP noise technology independent expert panel: Novel aircraft-noise technology review and medium- and long-term noise reduction goals. *ICAO Environment Documents*. Retrieved from https://www.icao.int/publications/Documents/10017_cons_en.pdf.

International Civil Aviation Organization (ICAO). (2014b). *Environmental technical manual, volume II, procedures for the emissions certification of aircraft engines (Doc 9501)*. International Civil Aviation Organization (ICAO).

International Civil Aviation Organization (ICAO). (2016, October). ICAO assembly resolution 39-1. *ICAO Environment Documents*. Retrieved from https://www.icao.int/environmental-protection/Documents/Resolution_A39_1.PDF.

International Coalition for Sustainable Aviation (ICSA). (2018). *Views on future supersonic transport engines and aircraft*. ICSA.

International Council on Clean Transportation (ICCT). (2018, October 2). U.S. passenger jets under ICAO's CO_2 standard, 2018–2038. *The ICCT*. Retrieved from https://www.theicct.org/sites/default/files/publications/Aircraft_CO2_Standard_US_20181002.pdf.

International Transport Forum (ITF). (2018). *(Un)certain skies? Drones in the world of tomorrow*. OECD.

Jakhu, S. (2011). *The need for an integrated regulatory regime for aviation and space*. New York: Springer.

Janek, J. (2016, September 8). A solid future for battery development. *Nature—Energy*. Retrieved from https://www.nature.com/articles/nenergy2016141#f1.

Kurzweil, R. (1999). *The Age of Spiritual Machines*. London: Phoenix.

Lawson, K. (2017, June 22). Drones for the transportation of CARGO. *AZO Robotics*. Retrieved from https://www.azorobotics.com/Article.aspx?ArticleID=226.

Leadbeater, C. (2016, November 17). London to Sydney in eight hours? How supersonic jets are set to transform travel. *The Telegraph UK*. Retrieved from https://www.telegraph.co.uk/travel/news/future-supersonic-flights-the-return-of-the-concorde/.

Ledgard, J. (2015, June 8). How to use drones for development. *World Economic Forum (WEF)*. Retrieved from https://www.weforum.org/agenda/2015/06/how-to-use-drones-for-development/.

Lilium. (2018). Simplicity was our most complicated goal. *Lilium—Technology*. Retrieved from https://lilium.com/technology/.

Lin, J. (2018, January 23). China's new drone company is building a UAV with a 20-ton payload. *Popular Science*. Retrieved from https://www.popsci.com/chinas-new-drone-company-has-big-plans/.

Lockheed Martin. (2017). X-59 QueSST. *Lockheed Martin Products*. Retrieved from https://lockheedmartin.com/en-us/products/quesst.html.

Matousek, M. (2018, April 29). Aviation companies are plotting the return of supersonic flight—And they think their jets will be better than the Concorde. *Business Insider UK*. Retrieved from https://www.businessinsider.com/aviation-companies-are-plotting-the-return-of-supersonic-flight-2018-4.

PwC. (2016, May). Clarity from above—PwC global report on the commercial applications of drone technology. *Drone Powered Solutions*. Retrieved from https://www.pwc.pl/en/drone-powered-solutions.html.

Red, C. (2008, December 18). Beyond the Concorde: Next-generation SSTs. *Composites World*. Retrieved from https://www.compositesworld.com/articles/beyond-the-concorde-next-generation-ssts.

Segarra, F. (2018, April 13). Drones: An artificial intelligence game changer. *CompTIA*. Retrieved from https://www.comptia.org/about-us/newsroom/blog/comptia-blog/2018/04/13/drones-an-artificial-intelligence-game-changer.

Simple Flying. (2019, February 7). Supersonic business travel coming soon? Boeing invests in Aerion. *Simple Flying*. Retrieved from https://simpleflying.com/boeing-supersonic-investment/.

Smith, K. (2017, November 7). This LA company is developing all-electric, zero-emission airplane. *Los Angeles Daily News*. Retrieved from https://www.dailynews.com/2017/11/07/this-la-company-is-developing-all-electric-zero-emission-airplane/.

Smout, A. (2017, September 27). Britain's easyJet aims for electric short-haul flights. *Reuters*. Retrieved from https://af.reuters.com/article/africaTech/idAFL8N1M82JU.

Sou, U. T. (2018). Electric aircraft and the environment: A literature review. *McNair Research Journal SJSU, 14*, 154–166.

Spike Aerospace. (n.d.). Quiet supersonic flight. *Spike Aerospace*. Retrieved on 24 June 2019, from https://www.spikeaerospace.com/s-512-supersonic-jet/quiet-supersonic-flight/.

Stolaroff, J. (2018). Energy use and life cycle greenhouse gas emissions of drones for commercial package delivery. *Nature Communications, 9*.

Swopes, B. (2017, October 8). Tag archives: Tissandier electric airship. *This Day in Aviation*. Retrieved from https://www.thisdayinaviation.com/tag/tissandier-electric-airship/.

University of Twente. (2018). Factsheet 1: Civil unmanned cargo aircraft. *The Platform for Unmanned Cargo Aircraft (PUCA)*. Retrieved on 14 August 2018, from https://www.platformuca.org/factsheet-1-civil-unmanned-cargo-aircraft/.

Warwick, G. (2012, November 2). Quiet supersonic—NASA chases fleeting booms. *Aviation Week*. Retrieved from http://m.aviationweek.com/blog/quiet-supersonic-nasa-chases-fleeting-booms.

Weber, A. (2017, January 5). Electric power will transform aviation. *Assembly Magazine*. Retrieved from https://www.assemblymag.com/articles/93676-electric-power-will-transform-aviation.

Wei-Haas, M. (2018, January 31). Inside the first solar-powered flight around the world. *Smithsonian*. Retrieved from https://www.smithsonianmag.com/innovation/inside-first-solar-powered-flight-around-world-180968000/.

Wey, C., Dutta, M., Patten, K., & Wuebbles, D. (2004). *Parametric analyses of potential effects on stratospheric and tropospheric ozone chemistry by a fleet of supersonic business jets projected in a 2020 atmosphere*. NASA Glen Research Center.

Wichmann, J. (2018, August 2). Our world is changing—But not as rapidly as people think. *World Economic Forum (WEF)*. Retrieved on 11 August 2018, from https://www.weforum.org/agenda/2018/08/change-is-not-accelerating-and-why-boring-companies-will-win/.

World Commission on Environment and Development (WCED). (1987). *Our common future*. Geneva: Oxford University Press.

The Yale Tribune. (2017, October 10). Fully automated: The disruptive technology reshaping the logistics industry. *The Yale Tribune—Tech, Politics and Everything in Between*. Retrieved from https://campuspress.yale.edu/tribune/fully-automated-the-disruptive-technology-reshaping-the-logistics-industry/.

Yenko, A. (2018, July 2). NASA's supersonic quiet aircraft for commercial travel gets a new name. *Tech Times*. Retrieved from https://www.techtimes.com/articles/231563/20180702/nasa-s-supersonic-quiet-aircraft-for-commercial-travel-gets-a-new-name.htm.

Index

T. Walker et al. (eds.), *Sustainable Aviation*,
https://doi.org/10.1007/978-3-030-28661-3

The manufacturer's authorised representative in the EU is Springer Nature Customer Service Centre GmbH, Europaplatz 3, 69115 Heidelberg, Germany. If you have any concerns regarding our products, please contact ProductSafety@springernature.com

Printed and bound by CPI Group (UK) Ltd, Croydon, CR0 4YY
30/09/2026
02225656-0013